Contemporary Irish Drama & Cultural Identity

Margaret Llewellyn Jones

First Published in Great Britain in Paperback in 2002 by
Intellect Books, PO Box 862, Bristol BS99 1DE, UK

Published in Paperback in USA in 2002 by
Intellect Ltd, ISBS, 5824 N.E. Hassalo St, Portland, Oregon 97213-3644, USA

Copy Editor: Lisa Morris
Typesetting: *Macstyle Ltd*, Scarborough, N. Yorkshire

Printed and bound by Antony Rowe Ltd, Eastbourne

A catalogue record for this book is available from the British Library

ISBN 1-84150-054-2

Dedication

In memory of my inspirational parents, Reg and Phyllis Scourse, and in celebration of their great-grandchildren.

Contents

Acknowledgements

In grateful appreciation of the following who gave interviews and information: Barabbas, Sebastian Barry, Mary Elizabeth Burke-Kennedy, Marina Carr, Darragh Carville, Anna Cutler, Seamus Deane, Dock Ward, Druid (Administration), Judy Friel (Abbey), Declan Gibbons (Macnas), David Grant (Lyric, Belfast), Pan Pan, Paul Mercier (Passion Machine), Paula McFetridge (Tinderbox), Lynne Parker (Rough Magic), Ridiculusmus, Karl Wallace (Kabosh), my son Edwyn Wilson (then Production Manager, Royal Court).

The Faculty Research Committee (Humanities and Education) and the Irish Studies Centre, both at the University of North London, provided welcome academic support, while my husband, actor Ray Llewellyn gave invaluable encouragement and Offstage Bookshop, London, obtained most playtexts.

1 Introduction: Staging Ireland

> The English did not invade Ireland – rather they seized a nearby island and invented the idea of Ireland. The notion 'Ireland' is largely a fiction created by the rulers of England in response to specific needs at a precise moment in British history.

Declan Kiberd's comment (1984, p. 5) reveals the crucial role that drama has played in the re-invention of Ireland. It challenges and subverts this British fiction through its exploration of cultural identity within theatrical space. Within the last ten years, especially, there has been a renaissance in Irish drama from both sides of the Border, including award-winning work that has transferred to London and New York, touring to Britain as well as Europe and Australia. As further discussed in this chapter, recent studies of postcolonial drama have tended to marginalise Ireland, while exploring non-European cultures. Strategies which have been associated with post-colonial theory and theatre practices; first theories of race, second varieties of feminism concerned with the representation of the gendered body, third roles of 'body, the voice, and the stage space as sites of resistance', and fourth varied use of cultural practices such as ritual and carnival (Gilbert & Tompkins 1996, p. 12) are present in much recent Irish drama. Intensified by the issue of language and its relation to power, these elements are all intrinsic to questions of Irish identity, representation and culture, and can be traced back to the link between the Literary Revival and the foundation of the Irish National Theatre. This book explores the dynamics of the relationship between these representations of Ireland and the fluid nature of cultural identity within a post-colonial context, especially during a period of economic and political change.

Unlike studies of Irish drama which have devoted chapters to individual writers, this book uses a thematic structure to reveal and analyse ways in which certain ideological concerns and dramatic strategies re-occur across a range of plays, in particular those features which may be linked to the post-colonial context. Although the book will provide a historical context for contemporary Irish drama, in order to establish these key themes it also includes discussion of some earlier works of the now canonical figures Brian Friel, Frank McGuinness and Tom Murphy, especially their more recent works from 1980 including some revived material. Well-documented writers from earlier periods, such as G. B. Shaw, Denis Johnston, Brendan Behan, and Samuel Beckett, are not explored here, since this book is a selective study which draws together more recent writers whose work has yet to receive wide critical attention within this particular context. The main emphasis lies upon works created by new writers performed during the 1990s and as they happen up to date, during the emergence of the 'Celtic Tiger Economy' in the Republic, and the continuing, if bumpy, Peace Process in the North. Selective reference is made to interviews with writers, performers, directors and groups, as well as performances seen in Dublin, Belfast, Galway and Britain, in main houses, fringe and community venues, since physicality of performance is central to the approach to these readings which are informed by aspects of post-colonial theory and perspectives on performance praxis, including feminism and psychoanalysis. Where references are made to plays encountered only as texts performative potential has been considered. Attention is paid not only to tensions associated with the colonising relationship between Britain and Ireland, but to the relationship between Ireland and Europe in terms of cultural and economic influences as well as performance practices, and that between Ireland and America in terms of 'the Dream of the West', the diaspora and tourism. Some allusions are also made to TV and film representations of Ireland and the Irish.

This Introductory chapter first provides a brief over-view of the historical and political condition of Ireland, second, a concise framework for the development of Irish drama through the Irish National Theatre and its relationship with issues of cultural identity. Third, with reference to the Irish context, it will comment in more detail on some key qualities and dramatic strategies that have been claimed to be typical of postcolonial writing and performance in general. Particular theoretical terms, key issues and themes that run through the book but also underpin its structure will be introduced. Finally, these aspects will be linked with the ways in which the especial use of realism in Irish drama is extended or breached, thus foregrounding the book's main case about the ideological significance of these representations for creating new and fluid cultural identities.

First then, a Time Line at the end of this book selectively indicates key moments both in Ireland's political history and theatre as a frame for reference, providing evidence of the colonial relationship with Britain. Whilst full justice to such a complex situation cannot be achieved diagrammatically, the Line does at least position theatrical events in context. It suggests how from the twelfth century England extended political and economic power over Ireland, particularly through the Tudor 'plantation' policy, which gave Irish estates to English gentlemen, thus eventually creating an Anglo-Irish Protestant Ascendancy ruling class. The Catholicism of the indigenous population provided further motivation for a series of Irish rebellions and excessively brutal English reprisals. Contemporary tensions between sectarian groups, deepened by immigration of Scots Presbyterians into the North, have seventeenth century roots, including Cromwellian massacres and the Battle of the Boyne. Social and economic repression was worsened by the application of the English Poor Law in 1838, and is most evident in the Great Famine period from 1845 to 1859, which depleted the population through death and mass emigration. It took over a hundred years from the Act of Union in 1800 for the English government to pass – but put on hold – a Home Rule Bill for Ireland. Following the Easter Rising of 1916, and the War of Independence, the Irish Free State was established in 1922, leaving the six predominantly Protestant counties of Northern Ireland as part of Britain. De Valera's constitutional claim to the more industrialised North in 1937 did not improve this problematic situation, which was exacerbated by the continuing of economic, social and political disadvantaging of the Catholic minority in Ulster.

Northern Ireland is still undergoing the last throes resulting from this colonial period, which were intensified by the decline of the crucial ship-building industry. A British province since 1921, it was virtually self-governing until 1972, despite economic difficulties rooted in inequalities linked to sectarian differences, which escalated from the mid-1960s. Direct Rule from London was reimposed after the fatalities of Bloody Sunday 1972 when British troops fired on unarmed civilian demonstrators. This period, 'the Troubles', was marked by Hunger Strikes, atrocities and reprisals from both sides of the sectarian divide in Northern Ireland, and included serious incidents on the English mainland. Agreements at Sunningdale (1973) and Hillsborough (1985) were attempts to find solutions such as a Power-Sharing Executive. On 10th April 1998 the signing of the Anglo-Irish Good Friday Agreement involving London, Dublin and Belfast, later supported by a referendum (May), granted Northern Ireland an Assembly and Cabinet at Stormont. Certain legislative powers were devolved from December 1999 (http://www.ni-assembly.gov.uk/about.htm). Amending Articles 2 & 3 of the Irish Constitution also relinquished the Republic's claim on Ulster (1999). Within an improving Northern economy, the Peace Process negotiations continued unsteadily during 2000 and 2001, variously challenged – for example by disruptive actions from both Nationalist and Unionist extremist splinter groups, and problems around de-commissioning of weapons.

Declared a Republic in 1948, Ireland, originally more rural, has become increasingly developed in terms of technology and industry, experiencing a so-called 'Tiger' economy during the 1990s. Financial growth – very visible in extensive re-building in central Dublin, where property values have increased drastically – is partly as a result of European Economic Community funding and also multinational investment, especially linked with American business. The Republic's participation in the European Monetary Union from January 1999 is a strong indicator of its present relationship in which Europe plays a more significant role than its past ambivalent links with Britain. It is confident enough to shrug off EU criticism about its policies of low taxation and expansion (*Observer,* 18 February 2001, pp. 16–7). *The Guardian* (26 August 2000) reports Philip Ryan, Deputy Director of the European Commission Offices in Dublin, as celebrating Ireland's self-confident identity – 'When Ireland joined the EU, 80% of our trade was with the UK, now it's about 27%'. Further, for the first time, migration back to Ireland, especially of young professionals, exceeds the number of those leaving, and successful Irish business is now investing in the regeneration of British cities such as Liverpool and Glasgow (*Observer* as above). There is an underbelly to this apparent prosperity, which includes not only poverty and drug problems but also corruption scandals, involving even allegations against Charlie Haughey, the previous Prime Minister. Economic development has changed the literal and social landscape, including attitudes to history, class and gender, and especially to religion, as the power of the Catholic church has declined. All these changing features can be traced in the nature and reception of dramas created on both sides of the Border, many of which revisit history, and can be linked to the evolution of notions of cultural identity.

Second, the origins of the relationship between drama and cultural identity are clear in Irish National Theatre history. In 1897 W. B. Yeats, Lady Gregory and Edward Martyn had produced a letter which they sent to a variety of prominent Irish men in the hope of gaining funds for founding a National Theatre. It became in effect a manifesto:

> We propose to have performed in Dublin in the spring of every year certain Celtic and Irish plays, which whatever be their degree of excellence be written with a high ambition, and so build up a Celtic and Irish school of dramatic literature. We hope to find in Ireland an uncorrupted and imaginative audience trained to listen by its passion for oratory, and believe that our desire to bring upon the stage the deeper thoughts and emotions of Ireland will ensure for us a tolerant welcome, and that freedom to experiment which is not found in the theatres of England. We will show that Ireland is not the home of buffoonery and easy sentiment, as it has been represented, but the home of ancient idealism.
>
> (Lady Gregory, 1973 edn., pp. 8–9)

These intentions – more problematic in practice than in expression – indicate the close relationship between the drive for national identity and cultural activity, also typical of the Gaelic League. Douglas Hyde, a League leader, for example, wrote and performed in *Casad-an tSugan* (the Twisting of the Rope) in Gaelic, for the Company in 1901. However, as Seamus Deane has indicated,

> The demands made by a dramatically broken history on writers who are caught between identities, Irish and British, Anglo Irish and Irish, Catholic and Protestant, are irresistable.
>
> (1987, pp. 15–16)

The fact that, however well-motivated, leading members associated with the National Theatre – including founders and writers – were from the privileged minority, the Protestant Ascendancy class, contributed towards various difficulties. Most notable were audience riots at performances of J. M. Synge's *Playboy of the Western World* in 1907 and Sean O'Casey's *The Plough and the Stars* in 1926. Both these plays, though in different ways, challenged idealistic notions of national identity. The new Irish National Theatre Society was born in 1902 from a conjunction of the Irish Literary Society founded by Yeats, Gregory, Martyn and George Moore in 1898, which had had production seasons running from 1899 to 1901, and the Irish National Dramatic Company run by the brothers Frank and William Fay, rooted in an amateur group of the 1890s. Following productions of plays in the Molesworth Hall, Dublin, in 1903 and early 1904, funding from the English eccentric Miss Annie Horniman enabled the new company to move into a permanent home, the Abbey Theatre. Within this new company there were tensions rooted in two significantly different approaches to production – Yeats and his associates tended to prioritise literary elements as a reaction against the shallow commercialism of English theatre which had invaded the Dublin stage, whereas the Fay brothers as actor/practitioners were concerned with employing and training Irish actors, particularly in clarity of speech, use of dialect and economy of action. Whereas Yeats disliked Ibsen for 'the stale odour of spilt poetry' (qtd. Flannery, 1989, p. 139), Frank Fay admired aspects of his social realism as well as his European theatre techniques. Ultimately, the Fay brothers resigned in 1908 under hostility from the rest of the Abbey management team and some of the actors, thus leaving the literary element dominant.

According to Christopher Murray, the Abbey's 'artistic standards floundered resoundingly' (1997, p. 142) during the late 1930s and 1940s, especially because of Ernest

Blythe's concern, in tune with de Valera's views, to prioritise work in the Irish language with little attention to the artistic value of those plays in English that were staged. During this period, according to Maxwell (1984, p. 136), such stultification can be linked to staging of romanticised views of peasant life in realist settings such as the middle room of a traditional three-roomed cottage, embodying a 'complacent shoddiness.' Roche also cites this deterioration, noting

> Rejection by the Abbey Theatre in general and Ernest Blythe in particular came to feature as a shared experience in the career of almost every contemporary playwright.
>
> (1994, p. 39)

The Gate Theatre, created in 1928 by Micheal MacLiammoir and Hilton Edwards, first rented the Peacock but opened at the Rotunda buildings in 1930. It saw national identity as only part of its concerns, originally intending to 'establish in Dublin an international theatre for the production of plays of unusual interest' and to attempt presentations precluded by commercial venues (Maxwell, 1984, p. 131). Problems linked with their colleague, Lord Edward Longford of Longford Productions, and the need to provide roles for MacLiammoir sometimes restricted this ambition between 1929 and the early 1960s. However, their 1929 production of Johnston's *The Old Lady Says No* was a hit, and in the immediate postwar years, Maura Laverty's plays, which exposed the horrific poverty of the Dublin poor were popular. The Pike theatre, which existed for nine years, was formed in 1953 to provide access to more innovative – and foreign – plays. It famously produced the world premiere of Behan's *The Quare Fellow* (1954) and the Irish premiere of Beckett's *Waiting for Godot* (1955). However, Murray, Roche and Grene (1999), in discussing the work of Behan, Beckett, Johnston and others from the 1950s, seem to agree that in the early 1960s there was a second renaissance of Irish drama. In 1964 the Gate's premiere of Friel's *Philadephia Here I Come* is considered by Richard Pine (1990, p. 1) as the start of contemporary Irish drama. Roche points out how fortunate Friel was to have Edwards, who had European experience, as his Director, deploying light, colour and humour effectively (1994, p. 85). The Gate has produced some works by Friel and McGuinness, but its programme still inclines to canonical plays from abroad, mixed with some from home, such as Joe Dowling's production of *Juno & the Paycock* in 1986. It has celebrated Beckett, stressing his role as an Irish writer in a comprehensive festival in 1991, which travelled world-wide finally showing at London's Barbican in 1999. Despite some subsidy for new writing, the Gate produces less new work than the Abbey – but examples include Dermot Bolger's *In High Germany* (1990), Joseph O'Connor's *The Weeping of Angels*, 1997, and Conor McPherson's *Port Authority* (with London's New Ambassadors) in 2001.

The Irish National Theatre at the Abbey was rebuilt in 1966 after a fire in 1951, with a mainhouse stage and the Peacock as a 'studio' space. The name 'Abbey' is used throughout this book as shorthand for the Irish National Theatre. An interview with Judy Friel (1997), then working as a Literary Manager at this theatre, suggested that subsidies for new plays tended to go to the Gate and the Abbey, which has now recovered a significant role including new writing. Schemes were in place for six-monthly 'Writers-in-Residence' at the Abbey through £7,000 sponsorships. New writing might be funded through commissions, or, when writers submitted synopses and scenes to the Literary Department, they might get the opportunity for 'inhouse' rehearsed readings, especially at lunchtimes. These sometimes proceed to full production. The image-rich work of Directors such as Patrick Mason has influenced the counterbalancing of the sovereignty of the word in productions. Distinctive new voices and issues are emerging – certainly the programme during one week in 1997 included an almost Jaques Lecoq-style re-working of Kavanagh's canonical rural novel, *Tarry Flynn* (1978), a cynical look at the underbelly of the Celtic Tiger economy in Jimmy Murphy's *A Picture of Paradise*, and a rehearsed reading of *A Different Rhyme*, an unsolicited script by Lorraine O'Brien. Further, work performed in the Abbey but written in the North, such as Friel's *The Freedom of the City* (1973), and more recently the premiere of Gary Mitchell's notable *In a Little World of Our Own* (1998) illustrates the change in previous tendencies noted by Murray for Southern audiences to 'detach themselves from the Northern troubles' (1997, p. 200). The Abbey's programme therefore includes production of many canonical and new writers discussed later under thematic chapter headings.

Many other central and peripherally-based companies and venues exist across the Republic and the North. Their work is acknowledged in this introduction and foregrounded in Chapter Seven. Murray cites the Ulster Literary Theatre, formed in 1902 with aims somewhat similar to those of the Abbey (1997, pp. 188–9), and the later evolution of the Group Theatre. The Lyric Theatre in Belfast, started informally by Mary O'Malley in 1951, gained a professional building in 1968. An interview with David Grant (then Assistant Artistic Director) indicates its situation in 1997, especially in relation to new writing (see Chapter Seven). The Field Day Company, discussed later in this introduction had a border-crossing significance that differs from the more centre-positioned Abbey.

The difference between Yeats and the Fay brothers, in their emphasis upon the verbal or the visual/physical, has been carried through from the Abbey's foundation into debates about the intrinsic quality of Irish theatre even into the 1980s and 1990s, which critics such as Nicholas Grene and Richard Kearney claim to have been primarily verbal until recently. Kearney refers to Thomas Kilroy's suggestion that

> the indigenous fascination with the play of language is a direct response to the Anglo-English experience of a displaced or de-centred cultural identity.
>
> (1988, p. 152)

Anglo-Irish playwright J. M. Synge's concern for Gaelic language and literary traditions which infused his plays with rhythms, vocabulary and structures respectfully drawn from – but not exact copies of – the speech of working people on the Aran islands and elsewhere, seems to have created a long term climate of expectation for audiences. For instance, despite Martin McDonagh's denial of such influences, his plays of the 1990s seem to draw upon 'received' notions of rural Irish dialogue. The notion that Hiberno-English is in itself a hybrid re-working of the coloniser's language, and thus a means of subverting it, could in some plays discussed later be linked with deconstructive theatrical strategies rather than recycled traditional verbal tropes. While, for example, it may be true that some contemporary writers such as Sebastian Barry write plays that are rich in poetry and evocative of the Irish canon, nevertheless his piece *Prayers of Sherkin*, first performed by the Abbey (at the Peacock) in November 1990 and revived by the Peter Hall Company at the Old Vic, London, in May 1997, draws upon visual images, body movement and stillness, with accompanying music and sound effects, as essential to its haunting, dramatic qualities. Further, as discussed later, even where some plays may still be apparently set in a conventional *mis en scene*, the frame of the fourth wall, conventional linguistic structures and playing styles increasingly draw upon physicality – for instance in ensemble work, directed by Patrick Mason for the Theatre of Images. Grene has suggested that Irish drama, through a 'crucial emphasis on its distinct form of Irish English', has with a few exceptions 'neglected or subordinated mis en scene' (1999, p. 268). Within the most apparently literary play, despite Grene's view, this book would suggest that the significance of the performing body in theatre space is crucial, particularly when one of the major concerns of Irish drama is the gap between what is said and what is done. As Eagleton puts it, this gap is 'the familiar Irish discrepancy between rhetoric and reality' (1995, p. 333), a borderline significant in colonial and post-colonial situation.

> The body cannot be thought of separately from the social formation, symbolic topography and the constitution of the subject. The body is neither a purely natural given nor is it merely a textual metaphor, it is a privileged operator for the transcoding of these other areas.
>
> (Stallybrass & White, 1986, p. 192)

Other aspects of the early Abbey programme that have continued to be significant are associated with tension between the mythic and the real, since this difference is relevant not only to content and style, but also to ideology. Current critical theory suggests that what has been defined as 'classic realism' is not ideologically innocent. According to Catherine Belsey (1980) and others, its apparent mirroring of the everyday world combined with an unproblematised presentation of a 'whole' human identity, with a linear narrative that proceeds from order to disorder to order, disguises its hierarchy of discourses and moves towards closure. These ideological perspectives, most evident in the closure of a realist text, are those in line with the values of the dominant group in a particular culture – in Western European terms, those of the bourgeoisie – and in a post-colonial context are thus likely to be problematic. Deane considers the Ascendancy's mythologising of the Celt was 'an attempt to reconcile on the level of myth what could not be reconciled in politics' (1987, p. 37). This ideological ambiguity is reflected in the original Abbey repertoire, which ranges between plays confined by a more realist structure set within fourth wall conventions and those which, in different ways, push against these limitations through magical/mythical elements, though not necessarily towards a fully subversive effect. This difference is indicated through a comparison between Yeats' *Cathleen Ni Houlihan* (1902) and his *The Countess Cathleen* (1899). In the first case myth expresses revolutionary spirit, and in the second re-affirms the spiritual virtue of the establishment. The movement towards Expressionism in O'Casey's plays, which challenge attitudes to Nationalism and social injustice, indicates his realisation that:

> Drama isn't going to stay quietly in the picture frame, gazing coyly out at the changing life around her, like a languid woman looking pensively out of a window in the fourth wall.
>
> (1985, p. 121).

A key feature of Irish contemporary drama is therefore both disruption of realist form and the re-working of mythic and folk elements as a means of deconstructing ideologies of language, history and gender through performance. This process informs the argument in the third section of this Introduction, and the Concluding chapter of this book. Most significantly, the relationship of Irish drama to European practitioners even from the Abbey's early days, and traced in detail by Worth (1986), persists to present-day experimentation. She traces the influences of Maeterlink and Lugne-Poe on Yeats' symbolism, that also drew upon aspects of Japanese Noh theatre. She claims that in many ways Yeats' use of dance, sound and physical movements anticipates the kind of approach to ritual and intercultural theatre practice that the ideas of Artaud (publ. 1964, 1981 reprint), Grotowski (1986 reprint) and Peter Brook (1980 reprint) have since brought to wider notice. Worth suggests that Synge's periods in Paris from 1895 also exposed him to the influence of Maeterlink and the symbolists so that he made 'rich Irish material [...] European' (1986, p. 139). In Paris, not only was Antoine's naturalism at the Theatre Libre (1887) influential, but Lugne-Poe's Theatre de L'Oeuvre's productions included Ibsen, Hauptmann and Strindberg. Yeats' reservations about Ibsenite realism were partly due to his own tendencies towards the mystical. Where Worth also discusses the Irish/French legacy of O'Casey's plays, Murray cites O'Casey's wide reading of modern European experimental dramatists (1997, p. 89). Yeats' famous rejection of O'Casey's *The Silver Tassie* in 1928 which precipitated his exile in Britain, did not preclude such European influences from gradually gaining strength in the Irish theatre. Roche has traced the ways in which Beckett, exiled in France, has strongly influenced generations of Irish writers from Behan onwards, but also mentions Devlin's *Ourselves Alone* (1985) as manifesting Chekhovian de-centred emphasis, politicised in Brechtian style (1994, p. 239). Murray acknowledges Behan's *The Hostage* (1958), extended to three Acts for Joan Littlewood's ground-breaking London production, as typically influenced by Bertholt Brecht's Epic theatre practice (1997, pp. 158–9). Throughout this book, European influences such as Artaud's notion of total theatre, Brecht's alienating Epic theatre techniques, as well as aspects of ritual and intercultural practice important to Artaud and to Brook (English, but based in Paris), will be shown as significant to ways in which realism is subverted or transcended, especially contributing to postcolonial dramatic strategies within contemporary Irish drama.

Further linked with the Abbey's past is the continuing problem of stereotypical representations. Yeats' fascination with Celtic mythology, the collection of folk tales by Lady Gregory and Synge as well as Synge's journal accounts of peasant life, somewhat challenge the notion of the Stage Irishman, who is strongly related to those stereotypical representations discussed by Roy Foster in *Paddy & Mr Punch* (1997). These show the Irish as brutalised and lazy, often with simian features. Eagleton explains the British susceptibility to stereotyping the Irish through its relationship with

> an island [...] unsettlingly close at hand [...] it is not with Ireland simply a question of some inscrutable Other, as an increasingly stereotypes discourse of stereotyping would have it; it is rather a conundrum of difference and identity in which Britain can never decide whether the Irish are their antithesis or mirror image, partner or parasite abortive offspring or sympathetic sibling.
>
> (1997 p. 127)

Typical crude representations persist, as in British television's *EastEnders'* soap opera visit to Ireland (September 1997), where basic stereotypes included large families, brow-beaten yet stoical mothers, peasant stupidity and trickery, drunken men, donkeys et al. This episode triggered hundreds of complaints to RTE (Radio Telefis Eireann), condemnation from the Irish Tourist Board, and was dubbed 'the curse of Paddywhackery in soaps' (*The Guardian* 26 September 1997). Ruth Barton's critique of the role of tourism and the heritage industry's role in perpetuating regressive media images of Irish identity points out that these indavertently function as postcolonial texts. (2000, p. 413)

When young and living in Tuam, a small town, dramatist Tom Murphy and his friend Noel O'Donoghue were determined that, whatever they wrote, 'One thing is fucking sure – its not going to be set in a kitchen'. (F. O'Toole, 1994, p. 22). Contemporary Irish drama's approach to stereotypical character and setting is, therefore, on the whole, deconstructive of role, location and audience expectation. For example, despite its title, *Kitchensink* by Paul Mercier, seen in England at the Tricycle, London (Passion Machine, March 1997), and set in a half built breeze block estate house, was cyclical in structure with archetypal overtones, performed by a cast who exchanged roles through stylised masks. In Declan Hughes' *Halloween Night,* directed by Lynne Parker (Rough Magic, April 1997), contemporary characters – friends, previous lovers and media folk – meet in a house on the sea's edge. They encounter mysterious events – including a possible revenant – and are threatened by a flood until huddling together, they mirror the desperate crowd on a raft in the vast reproduction hanging upstage. Plays explored in following chapters thus both carry further and subvert some of the Abbey's initial concerns as cited, through a developing challenge both to the primacy of language and to perceived stereotypes, especially in terms of gender, through disruption of realist form.

A major aspect of this book's analysis of Irish drama is the vexed question of its post-colonial status. Scott Brewster et al. indicate there are several ambiguities about the relationship between Britain and Ireland: 'the native Irish who were regarded as racially distinct, but were also white and difficult to place within the manichean dynamics of colonial racism' (1999, p. 2). Paradoxically, Kiberd pointed out, 'English rather than Gaelic became the language of Irish separatism and national politics' (1998, p. 12). Not only does geographical proximity within Europe and a shared language cause tension, but the identity question for those living north of the border is complex. As Anne

McClintock comments, 'for inhabitants of British-occupied Northern Ireland [...] there may be nothing 'post' about colonialism at all' (in Gibbons, 1996, p. 179). Just as this book does not separate plays into authorial 'clusters', neither are they grouped as 'Northern' or 'Republican' plays. This is not to evade the political issue of 'The Troubles' but to facilitate comparison of ways in which writers and companies have worked in particular though different contexts on similar themes. Differences in the cultural context – whether economic, religious or political – are discussed as and when they occur. Further, the recent burgeoning of drama in Ireland has not been separatist but almost entirely, in some way, a 'border-crossing' activity, contributing to the evolution of a more fluid, hybrid approach to cultural identity. The travelling of many of these plays to England and beyond indicates not only the presence of the diasporic Irish community, but also the potential for a more open and questioning attitude on behalf of British audiences than the attitudes recorded by Hickman and Walter in their 1997 Report on anti-Irish Discrimination for the Commission for Racial Equality. Audience responses are likely to differ according to both geographical location and communities of interest. For example, the iconographic significance of the donning of an Orange sash (see McGuinnesss *Observe the Sons of Ulster* 1985), or the presence of plastic statues of the Virgin Mary (McDonagh's *The Lonesome West* 1997), will evoke different and potentially challenging responses in the Republic, Ulster or Britain.

Events cited suggest Ireland as a whole is an appropriate subject of that criticism which analyses the long-term socio-political, economic and cultural effects of imperial power relationships, and is usually labelled as post-colonial or postcolonial. The hyphenated term 'post-colonial' indicates those matters which relate to the socio-political/economic structures and ideologies that underlie this condition, following the usage of theorists such as Aijaz Ahmad (1992). Since this book is mostly about textual products – plays and performances – created by writers living within this condition, here the term postcolonial, without the hyphen, is used most often, following the approach of theoreticians such as Homi Bhabha, whose works focus upon representation and discourse. The effects of colonisation can be seen in cultural products, but postcolonial discourses are closely bound up with the processes through which previously colonised peoples claim – and often re-work – their cultural identity.

As suggested previously, although some works recently published (and indicated in the Bibliography) discuss Irish drama as such, there has been relatively little extensive analysis of the most recent work as a postcolonial phenomenon, despite, the fact that, as Shaun Richards indicates in his Foreword to Brewster et al.'s *Ireland in Proximity*, 'The current (critical) location of Irish Studies is within the broad category of what is sometimes caustically referred to as 'pocopomo''(1999, p.xvi). Nevertheless, *The Post-colonial Studies Reader* (Ashcroft et al. 1995) finds room for only one article on Ireland (by Cairns and Richards) and does not even mention the country in its index. Articles and chapters within larger works have begun to engage with the subject from this perspective, and Richards himself is currently working on twentieth century drama. Lance Petttit's *Screening Ireland* (2000, p. 10ff) provides a comprehensive analysis of representations of Ireland in film and TV drama, framed by a useful discussion of critical debates about Ireland's postcolonial status. However, books by Crowe and Banfield

(1996), and Gilbert and Tompkins (1996), which centre on what they call 'post-colonial' drama, concentrate upon non-European playwrights. Nevertheless, the latter book in particular provides some useful guidelines about the potential nature of postcolonial drama, which can, with some reservations, mostly be applied to the Irish context.

The relevance of Gilbert and Tompkins' first concern with the 'intersection of dramatic theory with theories of race in a post-colonial context' (1996, p. 12) to Irish drama, has already been briefly indicated through reference to the Abbey's agenda, including the persistent problem of stereotyping as a representation of 'Other-ness', which runs through analysis in this book. Their second concern links the representation of the gendered body with Third-world feminism. However, in the case of Ireland, the designation here of separate Chapters (Four: *Madonna, Magdalen & Matriarch* and Five: *Myth & Masculinity*) to representations of women and men, foregrounds the importance of considering how the construction of both genders has been differently affected by the power of religion and the economic effects of the colonising relationship. For example, comparison of MacDonagh's *Leenane Trilogy* with Marie Jones' *Women on the Verge of HRT* and Marina Carr's *Portia Coughlan,* all seen during 1997, foregrounds dramatic strategies which, to different degrees, challenge and subvert culturally-based female stereotypes. Two male-written monologues also critique the oppression of women. The significant difference made by the rural or urban environment to representation of men is considered, for example, in Sebastian Barry's *Boss Grady's Boys* (1988) and J. Murphy's *Brother's of the Brush* (1993). Tracing the relationship to myths of heroism – whether Nationalist or Unionist – moves from Ron Hutchinson's *Rat in the Skull* (1984) and Graham Reid's *The Billy Plays* (1982/3) to Gary Mitchell's plays of the 1990s. Analysis also includes female-written plays about destructive patriarchs.

Gilbert and Tompkins' third, and most significant concern – the 'deployment of the body, the voice and of stage space as sites of resistance' (1996, p. 12) – is especially significant in the Irish context. While acknowledging the cultural importance of Irish landscape, this book suggests that performance space rather than geographical space is one within which the body may mark oppression or reclaim and recreate cultural identity/identities. Despite the role of verbal language as a marker of cultural identity, performance is created through a wide range of nonverbal signifying systems: lights, sound, costume, set and especially the actor's body in terms of kinesics (gesture) and proxemics (movements through space). Elizabeth Grosz suggests:

> The post-colonial body disrupts the constrained space and signification left to it by the colonisers and becomes a site for resistant inscription.
>
> (1990, in Gilbert & Tompkins, 1996, p. 204)

An example of this function from the early period is Lady Gregory's *The Rising of the Moon* (1905), where a subversive singer sits on the quay's edge bordering the audience – despite the 'Wanted' poster of himself behind him. Eventually both physically and verbally he out-manoeuvres a sergeant who is on watch. Similarly, during O'Casey's critique of Nationalism during 1915–1916 in *The Plough and the Stars* (1926), the physical proximity of Rosie Redmond, the prostitute, both to the Irish Citizen Army flag and the offstage rhetoric of Patrick Pearse, who was martyred in the Easter Rising, provoked riot through its deconstructive implications. Recently at London's Tricycle, the physical and narrative framing of events in Carr's *The Mai* (1997), by one of the Mai's daughters, drew the audience's attention to the way in which the eponymous heroine's behaviour was both like and unlike the heroines of myth – and traditional female stereotypes. For the acclaimed multi-authored *Convictions* (Belfast Festival, 2000, described fully in Chapter Seven), the audience was taken from space to space within what had been the Crumlin Road Courthouse. Being thus placed, their bodies could share something of the physical and emotional experiences both of the performers and previous inmates. Chapter Two: *Language & the Colonised Body* in particular introduces the performing body's function in the context of cultural identity, and in relation to the use of language. Ideas about the inscription upon the body of colonial power and the struggle against it will draw upon the theories of Foucault (1998) and Bakhtin (ed. Morris 1994). Analysis includes plays as different as Tom MacIntyre's *The Great Hunger* (1988) and Darragh Carville's *Language Roulette* (1996).

The role of (verbal) language both as a predominant signifier of Irish identity and in relation to physicality has been controversial since the establishment of the Abbey. It is through the acquisition of language that the human-being achieves identity. Through acquiescing in the speech contract, the individual enters not just the system of grammar but also the rules and ideology of a particular culture. Michel Foucault has pointed out that 'Discourse is the power to be seized' (1981 p. 53), so remaining silent creates powerful resistance. Since language is inevitably tainted with ideology, and is a means of maintaining hierarchies, the moment when a previously colonised country takes back the power to rule is often that when native language/languages become the official language – hence the aims of the Gaelic League and the emphasis on Gaelic/Irish within the Constitution of the Republic in 1922. However, by 1972 the Gaeltacht (Irish-speaking population) is reported to have declined to 32,000 residents and by 1973 compulsory qualifications were dropped from Leaving Certificate examinations and Civil Service entry (Foster, 1989, p. 596). In contemporary Ireland, the picture is complicated, as Tom Paulin indicates,

> State education in Northern Ireland is based upon a pragmatic view of the English language and a short-sighted assumption of colonial status, while education in the Irish Republic is based on an idealistic view of Irish which aims to conserve the language and assert the cultural difference of the country.
>
> (1986, p. 11)

His argument differentiates between the colourful and spontaneous qualities of spoken Irish English, and laments its lack of any institutional existence – which the Field Day project challenged through its dedication to using language as a means of social change. Brian Friel's play, *Translations* (1980), is particularly concerned with the function of language – its slipperiness, and the need to reshape/ reforge English in a way that not only looks back to the past but forward to the future. Drama embodies this 'challenge to persist in an aesthetic reconquest of that cultural self-image vanquished by the empirical fact of colonisation' (Kearney, 1988, p. 142).

Contemporary Irish plays therefore are likely to use 'voice' in a way that challenges past hierarchies through the use of Hiberno-English as a positive hybrid language. The plurality of Irish experience in the 1990s as opposed to the 1890s has developed due to social and economic changes this chapter acknowledges. Where Fintan O'Toole asserts, through quoting Leopold Bloom that 'A nation is the same people living in the same place [...] or also living in different places' (1994, pp. 12–13), Kiberd also stresses the tendency for Irish writers to write in a Babel of different languages. He lists a range of such authors who disprove the notion that great work is only possible in the mother-tongue (1995, p. 636). This hybridity is also claimed by Roche as intrinsic to the dual inheritance of Irish writers such as Samuel Beckett – 'the sense of English as somehow a language which is not entirely native' (1994, p. 4). Slippage between one tongue and another, or contrast between what is said and what is done, is often an expressive dramatic strategy. Later discussion includes not only some plays, which like Friel's *Translations* (1980) foreground language as a cultural issue, but also, as in Enda Walsh's *Disco Pigs* (1997/8), differently demonstrate postmodern qualities. Hybridity of voice is claimed as a further feature typical of postcolonial writers, who 'thus 're-inscribe the 'rhetoric', the heterogeneity of historical representation' (Gilbert & Tompkins, 1996, pp. 108–9), and is therefore especially significant in the Irish cultural context.

Since claiming geographical space is a prime imperialist strategy:

> There is a pressing need for the recovery of the land which because of the presence of the colonising outsider, is recoverable first only through the imagination [...] to the imagination of the anti-imperialist our space at home in the peripheries has been put to use by outsiders for their purpose.
>
> (Edward Said, 1988, pp. 11–12)

Extending Gilbert and Tompkins' views, this book suggests the use of theatrical space is especially significant in its concern with the imaginative representation of Irish cultural geography. Despite the traditional use of rural landscape and the valorisation of the family as a key indicator of the nation as fostered in the past by De Valera, economic changes have encouraged depopulation in the West and over-population and poverty in the more urban East of the Republic (Waters, 1997, qtd. Scott Brewster et al. 1999, p. 126). Throughout, but especially in Chapter Five, the significance of the growing representation of urban scenes and social problems is considered. The dominant, persistent problem of representing Ireland in rural rather than urban terms has tended to occlude not only the existence of such social problems, but also the double-sided nature both of the economic boom in the Republic and of political developments in the North. Further, the diasporic population implies that Ireland is both everywhere and nowhere – a notion that links psychological and geographical spaces around the world. Thus the relationship between the imaginative use of transformative theatre space and the creation of potential hybrid/fluid identities is discussed at length in Chapter Three *Politics, Memory & the Fractured Form*, which links Foucault's notion of heterotopia (1986) with Homi Bhabha's Third Space (1997). This chapter also considers the importance of space to the memory of history, and its relationship with mythic function and resistance to hierarchy, through reference to many writers including Frank McGuinness and Sebastian Barry, including detailed analysis of *Mutabilitie* (1998) and *The Steward of Christendom* (1995) respectively. The challenge to realism through post-Brechtian and post-Artaudian dramatic strategies includes the role of both split subjectivities and the revenant as a means of exploring the past, public or domestic.

Gilbert and Tompkins' fourth concern is the use of 'theatricalised cultural practices such as ritual and carnival to subvert imposed canonical traditions' (1996, p. 12). It is clear that traditional forms and practices, such as the seannachie (story-teller), the grotesque, the revenant, and the mixed/balancing of comedy and tragedy can be traced within many contemporary Irish plays – through certain cultural practices such as the Wake. Recent examples of the latter are Edna O'Brien's *Our Father*, premiered at the Almeida (2000) and Murphy's *The Wake* (1999), while the grotesque is central to Martin McDonagh's *A Skull in Connemara* (1997). Vivian Mercier (1991 reprint) and Toni O'Brien Johnson (1982) place these qualities firmly within Irish cultural tradition, but they can also be linked to Bakhtin's notions about the subversive carnivalesque ((ed.) Morris, P, 1994), as explained later, in the context of the relationship between body, performance and postcolonial theory. Further, myth and folk tradition play an important role not only through content which re-states a cultural heritage, but in those ritual actions performed on stage which invite a sense of audience involvement. While these latter elements may not be as strictly carnivalesque as the kind of total theatre strategies suggested by Artaud (1981 reprint), yet they are literally encoded within the performing body and may thus express postcolonial discourses. The later discussion of Friel's *Dancing at Lughnasa* (1990) and Murphy's *Famine* (1968), amongst others, support this critical approach which is also extended to other plays. The anti-canonical aspect of Gilbert and Tompkins' case, however, seems either more complex or less relevant, depending on whether the canon concerned is that of the Irish Literary Revival or the British Literary canon.

Usage of the term 'myth' here, extends from reference to Irish tales and traditions to embrace the position of Claude Levi-Strauss (1968). Put simply, Strauss suggests texts may have a 'mythic function' – that is to say they attempt to cover over or disguise paradoxical discrepancies and contradictions perceived in the empirical world. Fictions may be analysed in this sense through binary oppositions perceived, for example; good/bad, inside/outside, nature/culture, coloniser/colonised and so on. Space does not allow a full account of Strauss' complex tabular analyses, but the staging of McGuinness' *Mutabilitie* (see Chapter Three) and Murphy's *Famine* (see Chapter Two) demonstrates these oppositions physically. Since the mythic function of a text attempts to mediate between apparently unreconcilable positions, it may in itself bear the marks of this struggle in style and form, so Strauss' approach can be productively applied to postcolonial discourses. Kearney's analysis of Strauss states that 'myths are concerned with wish fulfilment and reversal, with making possible at an imaginary level what is impossible in our real or empirical experience' (1997, p. 109). In assessing the dangers of myth in the context of Nationalism, Kearney points out that it is a 'two-way street'. In postnationalist Ireland he sees the need both to de-mythologise and re-mythologise – but in a way which keeps 'mythological images in dialogue with history' (ibid., p. 121). This task is one that is carried out by many plays analysed later.

The general relationship between traditional elements within Irish drama and more contemporary representational strategies is complex in relation to history. According to Fintan O'Toole:

> What is peculiar about Ireland is that we have become a post-modern society without ever becoming a fully modern one.
>
> (1994, p. 35)

This leap is reflected in tension between the conflicting claims of the traditional and the post-modern, which provoke questions of identity. As Kearney indicates,

> The traditional ideology of cultural nationalism which claimed Gaelic to be Ireland's first language, Catholic its first religion and the re-unification of North and South its first political priority, has been increasingly eroded by the demands of present day reality.
>
> (1988, p. 1)

Furthermore he considers 'Both the 'unitarist' ideology of the South and the 'unionist' ideology of the North' (ibid., p. 10) are equally inappropriate. Significantly the consequent sense of fragmentation, discontinuity and pluralities of identity are embodied through a variety of experiments with dramatic form and style. Kearney classifies literary works within three categories. These are: 'revivalist modernism', which gravitates towards the past and tradition – for instance the influence of the Abbey; its polar opposite 'radicalist modernism' which resists the pull of tradition; and 'mediational modernism' – or perhaps 'postmodernism' – a 'collage of modern and traditional motifs', which in borrowing from both extremes is infused with a transitional or liminal/borderline quality (ibid., p. 15). This book considers the potential relationship of selected plays to these categories, with particular reference to the possibility that the middle 'third' way of mediational modernism may be the most common approach to the postcolonial condition. The notion of identity split between past and future is vividly presented by the split role of Private Gar /Public Gar in Brian Friel's *Philadelphia Here I Come* (1964). Unlike the characterisation of a realist text, Gar is presented problematically, in a way which interrogates the notion of human identity/subjectivity. As Gilbert and Tompkins suggest, 'Split or fragmented subjectivity reflects the many and often competing elements that define post-colonial identity' (1996, p. 23).

Other themes, dramatic structures and strategies common to many contemporary Irish plays can be linked with the post-colonial situation through the tension between tradition and modernism as defined by Kearney. These may include, for example: the function of verbal and oral language; the relationship of genre, ideology and form; the dynamic relationship of politics, history and myth; religion; perceptions of gender; traditional elements such as the grotesque; rural and urban contexts; diaspora, exile and return. These themes are expressed through re-working the conventions of realism by drawing upon the mythic, the grotesque and the carnivalesque body to question history, tradition, and national identity as a fixed, unified concept. Such strategies often also blur dramatic genres. The persistence of traditional tropes as guarantees of Irish authenticity has, for instance, been linked with 'Colonialism's initial denial of authenticity' (Graham & Kirkland, 1999, p. 15). Where revivalists tend to a backward glance, including the preservation of customs and ideas, Kearney suggests the Irish moderns tend to 'exploit rather than resolve the contemporary crisis of culture' (1988, pp. 12–13) and are prone to de-mythologize and interrogate tradition. This difference is closely linked to the extent to which writers have broken away from realist form, and its ideologically conservative closure, through dramatic and discursive strategies that may seem to be postmodern. Kearney's term 'mediational modernism' and its potential link with postmodernism is, however, indicative of another problematic – the relationship between postcolonial and postmodern discourses and representational strategies.

A major debate in this critical field is the nature of the relationship between the postcolonial and the postmodern. If history, language and the body are the main sites of ideological negotiation and post-colonial struggle, so these factors are particularly germane to contemporary Irish theatre practice. Its many deconstructive strategies such as disruption of linear time, fragmentation, splitting of the speaking subject, an emphasis of physicality and image, intertextuality, multiplicity rather than hierarchy of discourse, are typical of the postmodern. That is, the flow of signifying processes and their slippery nature produce a sense of 'surface', through exposing a proliferation of readings which undermine any notions of certainty or 'truth' held through the 'grand narratives' of the past. As Colin Counsell suggests, the postmodern artist is left with two options, one radical and one conservative: the former in its use of deconstructive strategies may 'provide an empowering understanding of cultural processes', thus laying bare their ideological function (1996, p. 205). The second option is to 'celebrate contemporary culture's celebration of surface and its resulting relativisation of meaning' which 'severs representation from reality, making art a separate realm' and operates according to the exchange values of contemporary capitalism (ibid.). Thus, although postcolonial discourse may share certain of the techniques and concerns of postmodernism, including the notion that identity is constructed, a major difference is its concern with history and political processes, including potential empowerment, more in tune with the radical approach cited. As Brydon reasons, their 'grids of interpretation' are different. Where 'the name 'post-modernism' suggests an aestheticizing of the political', she considers that 'the name 'post-colonialism' foregrounds the political as inevitably contaminating the aesthetic, but remaining distinguishable from it' (Ashcroft et al. 1995, p. 137). Through a critique of Linda Hutcheon's work on post-modernist and post-colonial practices, Brydon suggests that history is the nub of the difference. The former

> Without denying that things happened [...] focuses on the problems raised by history's textualised accessibility: on the problems of representation, and on the impossibility of retrieving truth. Post-colonialism, in contrast, without denying history's textualised accessibility, focuses on the reality of a past that has influenced the present.
>
> (ibid., p.142)

In contemporary Irish drama, although the actuality of the past and its effects is often evoked through postcolonial discourse, the problematics of contested representations of that past is often embedded in a form and structure which has postmodern elements. Plays discussed in Chapter Three: *Politics, Memory & the Fractured Form* are especially indicative of these aspects, including the difference between private memory and public events, which are embedded in forms and structures that have postmodern elements, as for example Vincent Wood's *At the Black Pig's Dyke* (1992) or McGuinness' *Mary & Lizzie* (1989).

Chapter Six: *Dreams & Diaspora: Exported Images & Multiple Identities* follows through this theme as it is centred upon ways in which the presence or absence of history in representations of Ireland that are marketed abroad operate to conservative or subversive effect. Analysis shows how generic form and structure refracts or reflects the proliferation of derogatory 'single' stereotypes – or may reveal the multiplicity of identities in the new Ireland. The roles of exile and revenant are especially linked to the relationship between Ireland and America, since a returning family member often acts as catalyst, challenging the cultural identities of those who have remained at home. This chapter combines a postmodern awareness of the constructedness of images and identity with a postcolonial concern about socio-economic factors that created both migration and the consequent ambivalent images of the country of origin, as indicated in plays about the diaspora. Images of identity constructed by the visiting film industry are epitomised in the contrast between McDonagh's *The Cripple of Inismaan* (1997) and Marie Jones' *Stones in His Pockets* (2000). The 'myths of the West' of Ireland and of America, as well as the role of tourism in the fabrication and marketing of idealised images, are explored in other plays with some brief reference to television drama.

In conclusion, Chapter Seven: *From Hearth to Heterotopia* draws together arguments about ways in which contemporary Irish drama has infiltrated the apparent realism of the hearthscene/fourth-wall theatre with subversive elements both postmodern and postcolonial, challenging dominant ideologies and creating a new 'heterotopic' space within which can be embodied the creation of new, fluid, intercultural identities. It reiterates the case for the especial importance of the physicality of the performer's body within space in postcolonial drama. It also acknowledges increasing involvement of Ireland with Europe, not only in the intertextual elements of the plays of Friel, Murphy and McGuinness, but especially those of newer writers such as Dermot Bolger, as well as the increasingly international profile of groups producing image-based work. Plays from both the Republic and Northern Ireland are discussed, which, perhaps in relation to the Peace Process, indicate new attitudes to the past and to issues of cultural identity, including gender and political perspectives. Examples include McGuinness' *Dolly West's Kitchen* and Mitchell's *The Force of Change*, both seen in London during 2000.

Although significant groups and venues are mentioned throughout this book and foregrounded in Chapter Seven, the Field Day Company merits special attention at this introductory point because its philosophy and practice to an extent set the climate and conditions for what has followed. In some ways a reaction to the centrally focused theatre of the Literary Revival and its aftermath, Field Day was originally founded in Derry, Northern Ireland, in 1980 by playwright Brian Friel and actor Stephen Rea, who were joined by poets Seamus Heaney and Tom Paulin, musician David Hammond and academic/creative writer Seamus Deane as directors. With a Board half Protestant and half Catholic, they obtained funding from the Arts Councils both in the North and the Republic. Like the Abbey in its concern with cultural identity, but critical of what has been called 'the calcified nationalism originating from the Irish Revival' (Pettit, 2000, p.17), the Company saw itself as operating within a post-colonial context. Field Day was very much involved with 'border-crossing', both literally on tour and conceptually in terms of plays produced. From 1980 to 1993, when it ceased such activities – partly through resignations prompted by Friel's problematic decision to stage *Dancing at Lughnasa* at the Abbey theatre – it produced twelve plays, some of which are analysed in this book. Deane and Rea were left as the only active members. Deane commented that in the colonial situation, 'What is robbed is the power of self-interpretation' (Interview, 1997). Therefore it was crucial to take back and reinvent the power of interpretation, 'not as an appendage of something called English literary tradition' (ibid.).

This concern with the discrepancy between language and actuality is evident in several of the dramas produced, supporting Deane's point that Field Day operated as a kind of

> early warning system [...] to liberate from these incarcerating languages in which we have been bred and find some new discourses which would enable people to think of different social arrangements – better and other than sectarian bitterness.
>
> (Interview)

Richtarik's extensive study of the Company from 1980 to 1984 indicates that it was 'attacked for being nationalist and for being 'anti-nationalist' (1994, p. 249). She also suggests Friel 'felt constrained by the ideological framework Field Day had developed by 1989' (ibid., p. 268). It is important to distinguish between the drama output of Field Day Theatre Company, and the considerable published output of Field Day itself as a whole (Murray, 1997, p. 209). These include fifteen influential pamphlets, published in three groups: *Series 1 & 2* (Derry 1983; London, Hutchinson 1985), *Series 3 & 4* (Derry 1985/6), and *Series 5* (Derry 1988; University of Minnesota, USA, 1989). Other published literary volumes reveal the group's wide range of topics of interest, broadly divisible into language/identity, history/mythology/vision (Murray 1997, p. 210) – yet suggest areas of potential disagreement. Other published works include the four-volume *Field Day Anthology of Irish Writing* (1991), the first three under Deane's general editorship. Pettit indicates the problematic reception of the volumes and their selection and omissions of writers as symptomatic of the debate on Ireland's post-colonial status. It revealed not only the persistent attachment of some critics to 'the idea of a nation' but also the search for an adequate means of representing 'Ireland's protracted crisis of identity', which Pettit attributes to Kearney's sense that the country is trapped in a 'transitional tension between revivalist and modernist perspectives' (Pettit, 2000, p. 17; Kearney, 1988, p. 10).

Field Day is doubly important. Firstly, for its emphasis on history, in which Deane agreed that pushing against realism is necessary,

> Historical divisions are not just made by accident, but made by someone. Historical conditions are created by the power system.
>
> (Interview)

and

> In order to express the experience of a post-colonial country, (one) must experiment with the formal norms of the dominant country.
>
> (ibid.)

Secondly, Field Day evokes the existence of a 'Fifth Province', a phrase appropriated from Kearney. According to Friel this

> [...] may well be a province of the mind, through which we hope to devise another way of looking at Ireland, or another possible Ireland, and this really is the pursuit of the company [...] Field Day is a forum where a more generous and noble notion of Irishness than the narrow inherited one can be discussed.
>
> (qtd Richtarik, 1994, p. 141)

The Fifth Province can therefore be seen as a liminal imaginative space, existing as it were between the existing political Borders, and which is evoked by theatre. It seems to have something in common with Foucault's heterotopia, and Bhabha's 'Third Space', discussed in more detail in Chapter Three. Thus the energy generated by Field Day persists.

As Judy Friel and others interviewed have stressed, in the Republic the benefit of tax-free earnings from their creative work, introduced for certain artists and writers by Charles Haughey in the 1970s has been productive. Further encouragement comes from Aosdana, an influential group to which significant writers belong, and from incentives such as the Stewart Parker Award, named after the Ulster playwright who died prematurely. During the last fifteen years throughout Ireland – and not only in the Festivals of Dublin, Galway and Belfast but throughout the provinces – a tremendous richness of different kinds of performance work has developed, much of it in fringe and community venues, often experimenting with increased physicality. Groups, writers and directors who are active in 'peripheral' as well as 'central' locations – to borrow Kiberd's (1995) terms – are mentioned throughout where thematically relevant: for example, Patrick Mason's work at the Abbey, Garry Hynes' with Druid Theatre, Lynne Parker's with Rough Magic, Marie Jones' with DubbelJoint and Charabanc. Companies such as Barrabas, Kabosh, Macnas, Pan Pan, Passion Machine, Ridiculismus, Storytellers and Tinderbox create transformative performances that continue to play an important part in celebrating Irish cultural identity locally, and on tour both at home and abroad. Venues such as the Gate, the Lyric and the Project Arts Centre in Dublin, the Lyric in Belfast, as well as smaller and community venues around the country offer a wide range of work. In London, venues such as the Royal Court, the Tricycle, the Gate, the Bush and sometimes the Donmar Warehouse put on Irish plays, particularly new ones, while the

National Theatre, Old Vic and Royal Shakespeare Company have produced works by successful writers such as Barry, McGuinness and McDonagh. Regretfully this relatively small book can only be selective, tending to refer to published texts, but acknowledges there is far more happening than it is possible to record. Although inevitably writing from an 'outside' perspective, the author's Welsh heritage at least suggests shared Celtic concerns. Hence a critical attitude to colonialism underlies the use of postcolonial theory, which is supported by broadly socialist and feminist perspectives. An attempt has been made to present an equivalent range of work from both Ulster and the Republic while stressing the border-crossing nature of many writers and companies, which is not to deny that audience responses may differ for ideological reasons among others. The author is appreciative of the co-operation of those interviewed, of productions seen – and of a culture that fascinates in its depth and development.

O'Toole states that whether at home or abroad, the Irish have an equal claim on the creation of Irish culture. Any mapping of the journeys of the Irish will depend on connections that are not physical but cultural, due to

> A sense of identity that is entirely imaginative, though not imaginary [...] cultural, matters not of a past that can be read but of a present and future that have to be constantly written and re-written.
>
> (1997, p. 179)

Contemporary Irish drama is in the process of such a recreation of identities.

2 Language and the Colonised Body

From Boucicault, through the Abbey repertoire, down to more recent films, landscape has been used as a key signifier of Irish nationhood, merging myth and notions of a lost heroic past. Eagleton has suggested that focusing on the aesthetics of a literary landscape can be conservative in its evasion of historic, economic and ideological issues (1995, p. 6). The dangers of romanticising Irish peasant life are captured in a note of J. M. Synge about the West,

> In a way it is all heartrending, in one place the people are starving, but wonderfully attractive and charming.
>
> ((ed.) Price, A, 1966, p. 283)

Through comparative though selective references to contemporary Irish dramas, this chapter will consider whether the presence of the language of the performing body may, in certain circumstances, encourage a more radical perspective. Argument extends the idea of mapping identity on to landscape by examining ways in which the performing body may be a privileged site, bearing traces of both culture and identity. The body may be seen as both a literal and metaphorical site of colonial power and the struggle against it. Postcolonial plays analysed seem, in accord with Kearney's comments on film, to operate,

> in assertion of the body as a site of political struggle where public and private coincide.
>
> ((ed.) Kenneally, 1992, p. 144)

Plays explored fall into three main categories: first where the relationship between body and landscape is associated mainly with social and economic restrictions; second those where the carnivalesque body embodies subversion of such boundaries; and third those in which the potential flexibility and slipperiness of spoken language is variously linked with bodily language in an urban setting. In all three categories, comparative analysis covers earlier and more recent work. Inevitably it includes the problematics of writing and performing history in the context of the pull towards tradition and authenticity, which Graham (1999, p. 15) analyses as a response to colonisation. However, these issues are discussed further in Chapter Three. Dramatic structures and strategies deployed include those previously introduced as typical of postcolonial drama and the different ways it disrupts traditional realist forms, with emphasis upon the significance of ritual. Foregrounding the body and challenging its limits is subversive, since the body is

> A model which can stand for any bounded system. Its boundaries can represent any boundaries that are threatened or precarious.
>
> (Mary Douglas 1966, p. 115).

In the case of Irish cultural identity any body-mapping inevitably relates to the scars of colonial and postcolonial ideology. Geographical violence cited as intrinsic to imperialism (Edward Said, 1988, p. 11) has in effect been writ large on the bodies of the colonised – as in the Great Famine – and it is through body mechanisms such as Hunger Strikes or Dirty Protests that twentieth century prisoners have attempted to subvert the coloniser's power. The body is perhaps even more significantly a bearer of meaning where the verbal language of the coloniser is still being used as the medium of drama. As the Introduction suggests, the use of a hybrid language such as Hiberno-English can have potential for subversion. This challenging quality may be enriched through kinesics (gesture) and proxemics (the spatial relationships between bodies and items on stage).

Clearly, productions vary in performance style and overall conception, while directors may choose to ignore certain stage directions or omit certain lines. Nevertheless, kinesics and proxemics are to a strong extent encoded within dramatic language, which differs from other kinds of literary language – as in novels or poems, for instance – through its use of deixis. Essentially deixis depends upon a more frequent use of parts of speech, such as adverbs of place and manner, which call for the 'intervention of the actor's body in the completion of (its) meaning', prioritising spatial over temporal relations (Elam, 1980, pp. 142–3). Speech Act theory can be used to analyse ways in which power relations are encoded in dramatic speech, such that characters are compelled to take certain actions or make certain verbal responses. More detailed expositions of deixis and of Speech Act theory can be found in Aston and Savona's *Theatre As Sign System* (1992, pp. 52–5). Because the written text is always a blueprint for production, analysis of plays below considers the performing body's role preferably through reference to productions seen. Otherwise discussion includes movement implied through the stage directions (nebentext), and/or the subtextual resonances provoked by deixis or speech acts embodied in the spoken text (haupttext). Further, other signifying systems – properties, lighting and setting – contribute not only to a production's atmosphere or the practical needs of the action, but also encode socio-economic status and symbolic qualities. In the case of postcolonial theatre, this symbolism is likely to embody culturally specific elements – some traditional – which will resonate more strongly with audiences from that culture. For example, personal experience of watching the same production of McDonagh's *The Lonesome West*, both in London and in Galway during 1997, emphasised this point, since the laughter patterns of the two audiences were different. Further, seminar discussions of Barry's *The Steward of Christendom* (1995) and McDonagh's *The Cripple of Inismaan* (1997) with both undergraduate and postgraduate students (in London) revealed differences in performance readings between those who were Irish (whether originating from Ulster or the Republic) and the rest, who came from many other cultures.

The geography of the performing body may well also encode the struggle to recover individual and cultural autonomy, especially where transgression of a borderline questions the 'limit rather than the identity of a culture' (Stallybrass & White, 1986, p. 200). Such liminality can also be associated with 'body borders', aspects that have been linked by Bakhtin ((ed.) Morris, 1984) in his theories about carnival and the grotesque, and further by Julia Kristeva to the abject body (1992). Just as the abject body may 'leak' through its physical apertures, contaminating other bodies, so experiences of carnival and ritual may be said to transgress the boundaries between spectator and spectacle in Artaudian style. Thus certain aspects of traditional ritual and other cultural practices, though differently drawn upon, may involve or 'contaminate' the audience through participation that may – even if only at a mental level, challenge or subvert dominant hierarchies. This kind of contamination supports Brydon's suggestion that postcolonialism inevitably contaminates the aesthetic by foregrounding the political (Ashcroft et al., 1995, p. 137). Differences in the plays' relation with realism may be related to Kearney's three categories of modernism (1988).

In the first category Brian Friel's *Translations* (1980) clearly links body, landscape and language to ideological issues associated with national identity, providing a useful starting point. As the first Field Day production, it was performed in Derry's Guildhall from 23rd September, 1980, then produced at the Abbey in 1983. Set in a hedge school in Baile Beag (Ballybeg), an Irish speaking community in County Donegal, during 1833, it dramatises a crucial moment in the process of colonial power, when Gaelic was being overtaken by English and enshrined through the Ordnance Survey mapping of the country. This task, carried out by the British Army Engineering Corps, anglicised many place-names as part of the integration of Ireland into the United Kingdom from 1800. Similarly, the native voluntary hedge-school is about to be replaced by a National School. Mapping/translation is shown as a strategy for fixing and controlling both physical and verbal discourse – and thus identity – within ideological limits. Hugh, an old teacher, ultimately grasps the paradox Friel himself has stated, that in order to reclaim Irish identity; 'We must make English identifiably our own language' (qtd. Pine, 1990, p. 155). Rather than nostalgically embrace a past rural idyll, the play exposes the problematics of communication endemic to verbal language of any kind, especially when tinged with cultural imperialism. Despite apparent focus upon spoken language, key moments foreground the language of the body deployed within theatre space, which like the exposure of the colonising project, accords with Gilbert and Tompkins' definitions of postcolonial theatre.

Hugh has two sons. Manus, lamed accidentally by his drunken father nevertheless helps him in the hedge school. Owen, returned from England, is interpreter to the mapping project run by the meticulous Captain Lancey, whose assistant Lieutenant Yolland describes him as 'the perfect colonial servant', influenced by possibilities triggered by the French Revolution of 1789. With a difference epitomised in his language and bearing, Yolland feels Ireland is a place where 'experience is of a totally different order' (1984, p. 416). This fascination deepens his friendship with Owen, but he falls in love with Maire, Manus' sweetheart, partly due to his over-romanticised and thus feminized view of Ireland. Although Hugh, the apparent protagonist, takes little part in the linear narrative, his view that 'it is not the literal past, the 'facts' of history, that shape us, but images of the past embodied in language' (ibid., p. 445) invites the audience to think beyond both the events and apparent reality of the stage, to question not only history but cartography. Throughout, the audience has to work with the double convention that most roles are spoken in Gaelic, and thus must imagine they are hearing two languages – an act that breaks with realism. All three Acts take place within the hedge school, set in a barn scattered with authentic implements and the remnants of cow stalls, their decay and the ramshackle tools sociometrically signify an outmoded and ineffective agriculture, and thus a disadvantaged community. Act 2 Scene 2, preferably through the use of lighting, happens 'down front, in a vaguely outside area' – a significantly liminal space where Maire and Yolland have their fleeting love scene (p. 426). Escaping from the dance into this magic circle of close proximity, they try to communicate. Maire tries out Latin she has learned in school, but Yolland mistakes it for Gaelic. She tries the elements – water, fire and then earth. Further misunderstanding follows with her sentence about Norfolk maypoles. Only when Yolland initiates a repetition of Irish place-names do they draw closer again, both speaking of each other's physicality, but neither understanding the other. Significantly, Yolland does not recognise 'always' among the words Maire speaks. Trembling, they kiss, observed by Sarah who loves Manus. After that moment of border-crossing, Yolland is no longer seen. His enthusiasm for land, language and love has put him in danger – he is probably killed by the Doalty brothers.

Throughout, the relationship between language, landscape and cultural identity is foregrounded through acts of naming and misrecognition. Cultural difference is heightened by the classical knowledge Hugh teaches, in which Jimmy Jack revels, but which is not understood by the British soldiers. Language's function as a prime element in identity is manifest in Manus' teaching Sarah, the dumb girl, to speak through naming herself. Ironically this skill later enables her to tell Manus about Yolland and Maire, thus precipitating his departure for Mayo. Sarah's muteness indicates her silenced status as both woman and colonized individual. With similar irony, when Maire evokes Daniel O'Connell's 'The old language is a barrier to progress' (1984, p. 400), her wish to learn English is not to reforge it within an Irish context, but to escape to America. The double-edged sword provided by knowledge of English is epitomised in Owen's changing attitude. Initially he translates Lancey's colonising strategies in a bland way (ibid., p. 406), but as his relationship with Yolland develops, a more equitable discussion of how place names may be adequately adapted emerges. Partly through Yolland's enthusiasm, and Hugh's significant attitude; 'It can happen that a civilisation can be imprisoned in a linguistic contour that not longer matches the language of … fact' (p. 419), Owen realises something more sinister than map-making is involved. Furthermore, Anglicised names can never capture their original cultural meaning. By the time Lancey reads out, in English, the names of hamlets to be destroyed if Yolland is not found, Owen translates accurately, and has to re-read the names in Gaelic. Realising the catalogue of names is merely a list – 'Nothing to do with us' – he goes off both to find the truth about Yolland from Doalty and to prepare to resist the evictions. Hugh as mediator claims the need to renew and rework images of the past for the future; having promised to teach Maire English, he remains with Jimmy Jack, whose passion is for past cultures. Despite its apparently realist structure, the unclosed ending leaves fluid not just immediate future events, but also the ongoing debate about the relationship of identity to language and land, which echoes through Field Day Pamphlets and beyond. Friel's diaries reveal that he was very anxious while writing *Translations*.

> Because the play has to do with language and only language. And if it becomes overwhelmed by that political element, it is lost.
>
> ((ed.) Murray, 1999, p. 75)

Despite Friel's anxieties, the dramatic strategies, which only slightly puncture the realist form through brief carnivalesque body-centred moments, nevertheless indicate a postmodern concern with the slipperiness of language, which cannot be separated from its ideological function in this context, where cultural identity is threatened by restrictions.

Two other early pieces also hinge on landscape and its relationship with ideologies that restrict the body. First, John B. Keane's *The Field* (1965) was staged in Dublin then New York (1976), and later made into a film (Granada Films International, 1990) that has since been shown on British TV. Second, based on Patrick Kavanagh's 1942 poem, is Tom Mac Intyre's similarly named stage adaptation *The Great Hunger* (1986) Both pieces evoke the harshness of rural life but are structurally and stylistically different, achieving different ideological effects. The former operates essentially as a classic realist text, the latter as an interrogative, postmodern piece in which an emphasis on the carnivalesque body is an agent of subversion. In both cases, the relationship of body and land is crucial to the representation of identity. Declan Kiberd's (1995) account of dire underdevelopment of Ireland's largely agricultural economy during the decolonizing period of the 1940s and 1950s contextualises such obsessive relationships with the land, the stifling impact of the patriarchal peasant community, and thwarted personal fulfilment evident in both plays.

The published version of *The Field* quoted here is the revised two-act production directed by Ben Barnes for the Abbey in 1987. Despite *The Field*'s popularity, comparison of its more conventional realist structure and performance style with both *The Great Hunger* and *Translations* suggests it is less challenging ideologically. Set in the 1930s within a small public house in Carraigthomond, a small village in south-west Ireland, its nebentext suggests fourth-wall realism in terms of properties used and actions indicated. A widow approaches Mick Flanagan, the landlord, about her need to sell a small field – 'Fine inchy grazing and as dry as a carpet' with access to water – which she had been leasing to Bull McCabe and his son Tadgh for grazing for the last five years. This pair intend to keep the field at a bargain price by auction-rigging, in tacit agreement with Mick and auctioneer Bird O'Donnell. When in Act 2 a stranger, William Dee, an Irish builder from London married to a relatively local woman, attempts to get the field with a much higher bid, both McCabes attack him in the dark when he goes to inspect the field. Although all the locals know who killed the stranger, and are in some ways also guilty through their tacit agreement to keep silent, they refuse to provide information to either the Sergeant or the Bishop. Even the widow is willing just to accept the McCabes' much lower payment for the field.

Landscape is shown as wide and harsh in the film version (1990), which begins with the McCabes throwing a dead donkey over a cliff. The play's simple pub setting provides an atmosphere of local claustrophobia. Only twice is this mimetic frame broken: first where the murderous beating occurs on the road, second when the Bishop seems to address both community and audience about the need to divulge information about the murderers. The first is a liminal space, indicative of the outsider's position relative to the rural community, the second a potential means of getting the audience to interrogate the moral issues. Although these are more openly discussed by Maimie Flanagan with her son Leamy, he is tempted to reveal the truth but emotionally blackmailed into keeping silent. Structurally, then, the form closes off any effective challenge to poverty's damaging effects, merely demonstrating a basic drive to survive through bodily violence. Verbal language also signifies this close relationship between land, body and survival, as in Bull's comments:

> I watched this field for forty years and my father before [...] every rib of grass; ' [...] I knew the wife was feeling the pinch lately [...] it was wrote as plain as a process across her forehead [...]
>
> (1990, p. 11)

Although Ben Barnes' production presented Bull less as a melodrama villain and more as a 'comprehensible individual', old and incapable of understanding social change (ibid., p. 8), male brutality remains dominant, underpinned by conflation of woman's body with property through humour – for example underlying sexual byplay with Maime Flanagan, 'a regular flier' (ibid., p. 121), and Dandy McCabe's joking 'auction' of his wife for her animal-like virtues (p. 134). The villagers' refusal to speak is not so much a means of resistance to forces of colonisation as in *Translations*, but a challenge to such economic changes – supported by the local hierarchy. Changes imposed on the land are in effect those achieved by 'strangers', who the locals feel want 'to bury (their) sweat and blood in concrete' (p. 136) yet herald greater social and industrial development. Ironically, the 'stranger' is, in the play, an Irishman – part of the diaspora – hoping to return home. Pettit's analysis of the film's differences (2000, pp. 124–6), stresses separation from the land as a marker of estrangement from cultural identity, as epitomised in the stranger as American, the widow selling the field as Anglo-Irish, and Bull's hostile attitude to the travellers. Pettit suggests the film's tragic element – Bull walking out to sea to drown after he has accidentally caused his son's death – re-works Yeat's version of the Cuchulainn myth. The play's narrower focus on rather regressive depictions of the villagers could evoke indignant responses similar to those provoked by Synge's *Playboy of the Western World* (1907). Thus the theatre version of *The Field* seems nearer to the revivalists, reflective of socio-economics rather than critical of postcolonial ideology.

The Great Hunger is a striking example of what has been called The Theatre of Images, where not only silence, but also the body constantly provide sites of resistance; thus it falls within the second category cited in this chapter's second paragraph. Originally performed at the Peacock in 1983, then successfully taken both to the Edinburgh Festival and London's Almeida theatre in 1986, it is non-linear in approach, with unsettling and surreal images predominating over an unusual use of verbal language. Essentially collaborative, originating in a meeting between writer MacIntyre, director Patrick Mason and principal actor Tom Hickey in 1983, a draft script grew into a first rehearsal script, developing further from company rehearsals. This imagistic style – somewhat evocative of European film directors Bunuel, Cocteau and Fellini in its cinematic quality – has since evolved further through Mason's approach. In terms of theatre practice, the physical style resonates with practitioners Grotowski, Brook, Eugenio Barba and more recently Jacques Lecoq in Europe. This Dublin venture both demonstrates the European outlook of Irish theatre and anticipates the rich development of Irish physical/performance art/avant garde work described in Chapter Seven. Production aspects cannot be easily recaptured in words, but the published text contains a very detailed nebentext, clearly indicative of striking emotional responses triggered, and positive responses from critics as diverse as Michael Billington and John Barber (1988, p. 30). Lasting around one hundred minutes, ideally without an interval, six players, three male and three female perform individual roles and a host of other figures, moving rapidly and flexibly across a loosely defined fluid space. Outdoors is signified centrally, with a wooden gate far upstage. Downstage left and right are kitchen, indicated by a large black kettle and bucket, and chapel suggested by a tabernacle resting on its pedestal. At the Almeida, the upstage back wall consisted of a towering wall of blackened and rusty zinc against which stones were thrown, while the centre of the field had potato-field ridges (Etherton, 1989, p. 46). Typical of the play's deconstructive mode is the representation of Mother by a wooden effigy – usually found in the kitchen. This collapsing of religious and domestic restrictions into one Madonna-like sign brilliantly indicates gender stereotypes discussed in Chapter 4. Sparse properties convey the struggle for existence typical of the stony grey hill soil or boggy low terrain of Cavan-Monaghan, associated with Kavanagh and his work.

The character Maguire is used to reveal the crushing effect of rural labour, limited community, repressed sexuality, and particularly the Church's restrictive power, during the early 1940s when Ireland was neutral during the Second World War. The title implies this desolate hunger for wider spiritual and sexual experience is as serious as that caused by lack of food in the Great Famine. Although not split into two roles (Maguire and Maguire Poet), as originally intended during the rehearsal/devising process, Maguire sometimes comments objectively on his activities:

> Patrick Maguire went home and made cocoa […] The sister, the sister – hens and calves, calves and hens.
>
> (1988, p. 37, p. 50)

which gives an impression of a self divided between mundane routine and poetic yearning

> The bridge is too narrow […]
>
> (ibid., p. 35)

Veering between close involvement in activities and framing observation of others – for example when voyeuristically watching Agnes flirting with Malone – Maguire eventually hangs upside down on the farside of the gate, looking through prison-like bars (p. 60). Indecision about his own identity is highlighted as he looks at his face in the door/mirror (p. 56) and stares speechless and desolate at the audience. Moments of stillness alternate with moments of frantic activity, juxtaposing the poetic and the grotesque. For example, Maguire's frustrated sexuality, indicated when his act of blowing on the fire to keep it alight accelerates into masturbation (p. 39, p. 50).

Image sequences convey aspects of Maguire's routine at home, field and church, including tantalising moments with village girls or observing their flirtations, until soon after his mother's funeral, drunk or dying he lies on and seems to merge with the ground. Complex physical routines fall loosely into two categories. The first embodies the sheer slog of potato picking (1988, p. 36), domestic cleaning (pp. 38–9), or the exhilaration of ploughing (pp. 52–3). The second energetically subverts or diverts these tasks – for example by playing pitch and toss (pp. 47–9) or a version of maypole dancing (p. 58) and other games of sexual byplay (p. 46). Actors transform themselves through physicality, as in the Heifer romp (pp. 50–1). A third kind of ritualised group movement is more transgressive, subverting religious practices and the revered status of the Mother/Madonna figure, while evoking Christianity and paganism through waving green branches:

> *The green branch is magic. For each an individual way of dealing with it. Maguire is ecstatic, Mary Anne severe [...] Agnes lies down [...] strokes the branch against thighs, breasts, face [...] Packy's grotesque gestures convey delight [...]*
>
> *SCHOOL GIRL Holy Spirit is the rising sap [...] Holy Spirit is the rising sap [...]*
>
> (pp. 42–5)

Interludes involving choreographed waving of church collection boxes are more overtly satirical, an irony enhanced by the priest who, although offering the host, elsewhere performs other magic, that is, conjuring tricks. In the first type of movement sheer physical energy illuminates the nature of labour, in the second it transforms this into play to an almost frenzied degree, indicative of the repressed sexuality and imaginative potential contained by the land's rigorous demands. As Stallybrass and White have observed about the nature of hysteria:

> It is striking how the thematics of carnival pleasure – eating, inversion, dirt, sex, stylised body movements – find their neurasthenic, unstable and mimicked counterparts in the discourse of hysteria.
>
> (1986, p. 182)

Various games involve dirt, eating and sexual gestures that evoke human orifices and the abject; animal interludes draw further attention to the slippage between the borders human/animal, and human/spiritual associated with carnival celebration. Just as rituals

> enact the form of social relations and in giving these relations visible expression they enable people to know their own society. They work upon the body politic too through the symbolic medium of the physical body.
>
> (Douglas, M., 1966/1984, p. 128)

Hence this third category of movement deconstructs church ritual, challenging its power. The conflation of Madonna/Mother in procession ritual, and the role of Mother as effigy within the home likewise demonstrates the stultifying power of the State's version of family life. Yet, as the priest says, 'Your children will miss you when you're gone' (1988, p. 40), and ultimately Maguire shows terror of Mother's corpse and the possibility of life without her, despite her association with the bag/apron of tools and tasks that restrict him. These subversive attitudes to gender and religious ritual, both embedded in Irish culture, are typical of postcolonial dramatic strategies.

This drama's postcolonial ideology is enhanced by its postmodern structure, which leaves it open for a plurality of readings further strengthened by language usage that disrupts traditional forms. Repetitive structures accompany and enhance physical rhythms, as in the opening potato picking scene,

> Move forward the basket –
> The wind's over Brannigan's –
> Balance it steady –
> That means rain –
> Down the ruckety pass –
> The wind's over Brannigan's –
> That means rain [...]
>
> (1988, p. 36)

Surreal antiphonal and polyphonic passages blur the identity of individual speakers, as between Priest and congregation,

> Remember Eileen Farrelly? I was thinking a man could do a damn sight worse. She ought to give a crop if any land gives [...]
>
> (p. 43)

Such mingled phrases, often evoking the land, are accompanied by non-verbal sounds and animal grunts or seagull shrieks – while in a small venue the sheer sound of energetic actors' breathing would be evident. Similar fragmentation of speech accompanies physical interchange in a group game querying the mystery of the identification of God with a tree, a crumb of bread or green leaves. The primacy of images and merged sounds evoke polymorphous sensations, which psychoanalytic critics have associated with the experience of the pre-speech child before the splitting of the human subject into the speaking and the unconscious self (Kristeva 1986, Jacques Lacan 1977). Embodied in a mediational modernist performance, which explores the problem of establishing individual identity within a restrictive rural landscape, these imaginative dramatic techniques demand a highly integrated and skilful ensemble company. Audiences can respond not only to the authenticity of tasks shown but to the power of ritualised images within deconstructive visual and verbal postcolonial discourse.

Where both *The Field* and *The Great Hunger* differently explore spiritual as well as economic deprivation without aestheticizing the hardship involved, Tom Murphy's *Famine* faces the problem of representing the Great Hunger, that is, the potato famine of the 1840s. In calling this terrible event 'the Irish Auschwitz', Eagleton comments 'there would seem something trivializing or dangerously familiarising about the very act of representation itself' (1995, p. 13). Murphy's success can be indicated partly in comparison with the strategies of the television serial *The Hanging Gale* (1995), based on an original idea by Stephen and Joseph McGann, whose family left Ireland after the famine. Made by Little Bird (Ulster) it was the largest production to that date for BBC Northern Ireland in association with Radio Telefis Eireann, with support from the Irish Film Board. Murphy's play was directed by Tomas Mac Anna at the Peacock in 1968. It was very successfully revived in a large-scale production by Druid Theatre Company during 1984, using a large ballroom in Salthill rather than their small Galway theatre space, prior to a tour.

Famine is set in one community, Glanconor. Its moral centre is the contradictory qualities of their reluctant leader, a man who comes to draw upon what Murphy calls his inner or 'sacred strength', holding what is essentially a pacifist position, while not being in himself a conventionally moral man (1992, p. xvi). As the place name suggests, Connor's ancestors were once local princes and there are mythic/archetypal undertones in his relationship to both family and community. Believing that no play can do justice to famine, Murphy has deliberately set the play to conclude in the spring of 1847, thus not reaching the worst period of 'Black 1847' (p. xvii). O'Toole's assessment of Murphy's oeuvre claims the increasing disjunction between private and public values shown,

> disintegration of shared traditions and values, the common ideology of Catholic and Nationalist Ireland, and their replacement by an individualist ethic.
>
> (1994, p. 124)

is similar to the erosion in the 1960s of those 'ideals of duty, shared notions of what is right, the appeal to the authority of the past' as shown by Connor, but replaced during the late twentieth century by individual, consumer-led priorities.

It is impossible to give a brief uncontested summary of causes and events of the Irish Famine. Woodham Smith's classic account of this disaster, *The Great Hunger 1845–1859* (1964), has been superceded by others, which by no means mutually agree. Baldly, in a colonial situation the poverty of Ireland and the importance of the potato to the rural population, whose land had frequently been subdivided among often large families over generations, were factors made fatal by the invasion of the particular potato-killing fungus *Phytophtora infestans*:

> Despite the massive decline in availability of food, the huge death toll of one million (from a population of 8.5 million) was hardly inevitable; there are grounds for supporting the view that a less doctrinaire attitude to famine relief would have saved many lives.
>
> (O'Grada, reprint 1989, back cover)

Indifferent and slow reaction by the British government in particular, linking of relief to schemes such as road-making, as well as dubious behaviour by some landowners (absentee or otherwise) and by indigenous bourgeois' commercial trading of other foodstuffs were among the complex factors which have been considered as partly responsible. The church's role was also ambiguous. By any standards this nineteenth century catastrophe, which also produced mass emigration, is a turning point in modern Irish history.

Beginning with funeral rites for Connor's daughter, the play includes attempts to forestall the carts of corn en route for England, struggles with police, priest and landlord's agent, Connor's attempt to hold family and community together as hunger provokes despair, and different views of what should be done. In the face of threatened supported emigration, an excess of deaths that prevents dignified burials, futile road labouring schemes and evictions, Connor's starving wife incites him to murder her and their youngest child. Connor, perhaps mad, is left with a much depleted community, including Maeve, her lover, and Mickeleen, a crippled renegade.

Formally *Famine* is fragmented and episodic, pinpointing key issues so that, despite relatively rounded main characters, the piece has a somewhat Brechtian flavour, with carnivalesque touches in which ritualised elements emotionally draw in the audience. A wide range of views expressed in the community avoids providing a hierarchy of discourse, but epitomise the crisis. Carefully researched historical details support Murphy's overt concern with the other poverties that, as well as the absence of food, attend famine and erode humanity. Key counterpointed images embody these absences (1992, p. xv). Stagecraft that foregrounds the body crystallises various contradictory positions from the opening, when grubbing up of bad potatoes erupts into a carnivalesque chase and a dance, which contrasts with the keening Mother's quiet dignity. Her ritualised words and the group's reiteration of phrases from her prayer as the coffin is closed express intrinsic cultural tradition. The potato throwers and the Mother thus embody different challenges to the colonising hierarchies' ethos. These intensify as the regular rhythm of corn carts offstage is set against Connor's attempt to prevent Mickeleen and his brother Malachy seizing the carts, and the ensuing fight when Priest Horan incites the crowd to turn on the cripple for his past acceptance of Protestant soup. Proxemic relationships embody the paradoxical binary oppositions resulting from famine. For example, a graphic image from Scene 4 juxtaposes the corpses of a woman, her two children and the body of a dying man on one side of the stage, with a developing love scene between Maeve and Liam, who share an apple and nuts which he has buried, on the other side. Mother's more pragmatic approach when she is reluctant to feed those meeting at Connor's house before her family, is contrasted with his belief in rules and hope in God's help, as she stirs a pot, upstage. Once Connor has refused to agree to emigrate, they are dispossessed and starving. Later, on one side of the stage Connor sits motionless in the ruin of his house, while on the other Dan, in a mixture of madness and keening, remembers the past over his wife's corpse, thus demonstrating great events of history mean little to the marginalised and powerless. Ultimately, Mother goads Connor into taking the freedom to kill her and her young son 'lest they (ie: Others) choose the time and have the victory' (1992, p. 88). The sound of his beating stick offstage signifies their death – the ironic result of idealism, and a blow to the sacred image of the ideal family. These events, like the representation of famine through ritualised techniques, have some overtones of European practitioner Artaud's 'Theatre of Cruelty', since although the play does not fully break down barriers between spectator and audience, the horror presented has a shock potential that can stir the audience to 'a re-examination not only of all aspects of an objective, descriptive outside world but also all aspects of the inner world [...]' (1981 reprint, p. 71).

Despite poetic passages, such as the opening keening, Dan's mourning speech or parts of the ironically titled love interlude, both verbal and visual language is grotesque and harsh. Crippled Mickeleen owes his grotesque deformity not to nature, but to thrashings from his father's stick. Even Connor is shown harshly ignoring his heir Donnail, and using him to test out a grotesque reusable coffin in a tragi-comic scene. Passages reveal the double-think of those in power who evaded their responsibilities, showing their racist perspectives – particularly the Epic theatre style Relief Committee scene in which no villagers appear:

> CAPTAIN: Is there a pure Irish race somewhere – or are you referring to the monkeys roaming the hills [...] who could now be men, but for Popery that keeps them apart [...]
>
> (1992, p. 52)

> CAPTAIN: Did you ever try to get an honest day's work out of one of them [...] would you expect it of a black man? Ignorance, deceit, rent evasion, begging [...] Filth, the breeding of disease [...]
>
> (ibid., p. 51)

Unlike the more complex depiction of the main characters, figures such as the Landlord, Captain and Policeman are portrayed as gestic examples of their social function in the Brechtian manner, and their dialogue is less motivated. Rather than accept the responsibility of absentee British Landlords, the Agent tries to blame Irish MPs who supported Daniel O'Connell (the Liberator), attributing to them the

> belief that if the people starve patiently, the result will be a speedier repeal of the Union between Great Britain and Ireland
>
> (p. 55)

Although Murphy is critical of the church's role, nevertheless the Parish Priest attributes the people's misery directly to 'Colonization and poverty' (p. 57).

An open, interrogative end can draw upon audience knowledge that even worse famine succeeded this episode. Although more dashing brother Malachy has escaped to America, grotesque Mickeleen still lives. Connor sits distantly upstage and silent. The arrival of bread is offset by the presence of a corpse, while potential romance between Liam and Maeve has been damaged by events. This downbeat ending is ambivalently expressed in Maeve's response: 'There's nothing of goodness or kindness in the world for anyone, but we'll be equal to it yet' (ibid., p. 89). Left to question not only the range of positions shown, the audience might, for instance, consider previous Irish President Mary Robinson's heavy involvement in encouraging support for the Third World.

Paradoxically, Murphy's deployment of the body on stage emphasises its abject qualities through the inscription of hunger, violence and death. Nevertheless there are moments when carnival energy also reveals the positive dynamic that has been repressed by economic and colonial restrictions, such that it has been compelled to burst forth in a kind of freedom in revolt and murder. Reworked traditional elements blended with interrogative modernist techniques infused with comedy, balance the predominantly tragic tone.

In contrast, *The Hanging Gale*'s representation of the famine runs in some danger of the film director's dilemma of

> turning abject poverty itself, by handling it in a modish, technically perfect way, into an object of enjoyment.
>
> (W. Benjamin, 1973, pp. 94–5)

Irish landscape, which has often elsewhere upstaged events and characters, is shot here in a way that veers between realism and romanticism. As Gibbons has suggested, these polarised approaches reflect tension between the perspective of a dispassionate outsider and that of either a tourist or a coloniser (1996). More conventionally structured episodes follow classic realist narrative format, with associated limitations of unproblematised characterization, ideological hierarchy and closure which tends to resolve contradictions. Nevertheless, certain strategies sometimes subvert these limitations, while retaining some elements of the Victorian melodrama tradition Rockett associates with Irish popular culture (1987, p. xiii).

The four Liverpudlian McGann brothers feature as the beleaguered Phelan family, related to the Dolans by marriage. Set against them is the Agent, an ex-Indian Army Officer, come to Galreedy to represent Lord Hawkesborough, an absentee landlord, as replacement for an Agent whose brutal murder (by local men in disguise) starts the tale. Often shot from a low angle, ironically riding the white (grey) horse of chivalry, Townsend is a guilty figure torn between his own need for survival, the blindness of his absentee master and sometimes sneaking sympathy for the peasants – victims of hunger and colonial brutality, including his own. Escaping a series of attacks, he is killed once he stops making concessions, although not before an abortive rape attempt upon Mary Dolan, who becomes his servant after her parent's eviction. As a colonising Englishman, his threatened and damaged body shows the dangers of border-crossing or falling between the limits of power and the needs of humanity.

While as central symbol, the suffering and starving Irish body is not gratuitously foregrounded, one scene shows the priest finding one dead and one dying child in a tiny shelter in wild country. Rather, sharp cutting between sparse cot and the Agent's house exposes the gap between colonised and coloniser. Within the linear thread, diaries and letters of protest from Doctor, Priest and Agent as well as denials from England are shown either being written or read. Their contents sometimes operate as voice-over. These instances of metanarrative often drawn from historical sources, ironically reveal the gap between what is said and seen, or indicate the process of history construction. Characterisation is compounded by the brothers' roles as mouthpieces for different political discourses: priest, Ribbon Man and Teacher, husband and father, loving yet pugilist brother. Despite this attempt at representing diverse ideological perspectives, subsuming of them all under one – loyalty to the family – reinstates a traditional hierarchy of discourse. Whereas in *Famine* the idealistic Father is a major instrument of the family's fate and ultimately a murderer, here the Father's suicide and family breakdown is shown to be largely due to external forces. Class exploitation is shown through the unscrupulous provisioner Coulter, who also leads the anti-British anti-landlord group. Despite the use of historical sources, including Woodham-Smith, the extent of British responsibility seems understated.

Family melodrama stereotypes limit characterisation: both Phelan and Dolan patriarchs are – though differently – weak and wrong-headed. Both young women are potential sites for positive and disruptive powers of desire and sexuality. Yet Maeve, the young wife, has traditional strength and virtues. Her first Madonna-like appearance in the film is iconic. Head covered, carrying her baby and potatoes for the men's lunch, she is framed against breathtaking sea and landscape. Mary, spurned by the rest of her family for seeking survival in the Agent's house, succumbs to the Ribbon Man she loves, but is then unfairly rejected by him as damaged goods. Thus the female body is here measured by the typical Madonna/Magdalen polarity discussed in Chapter Four, which Murphy's more complex treatment avoids.

Where Murphy's play, through economical yet vivid staging, conveys the situation's horror, lingering landscape shots of *The Hanging Gale* do at moments verge on visual excess. Despite deconstructive juxtapositions and some evidence of muddy agricultural effort, views of humble interiors take on the tonal texture of Old Masters. Heroic frames of the brothers intertextually echo pioneering Westerns, while romantic interludes are almost strong enough to outweigh the central historic injustice's traumatic power. Further, closure reasserts the life of the family after the death of two brothers; first through news that Mary has borne one's illegitimate son, and second because Maeve, the other's widow, marries the remaining brother on emigration. Rather than a downbeat or radical ending, this confirms Hollywood mythology. *Famine*'s strategies of interrogation seem to offer a way out of such an impasse borne of myth and romanticism by opening a gap between image and history through performing the suffering of the Irish body denied by colonial rule. *The Hanging Gale*'s predominantly realist approach recognises the body's hunger as a defining limit of cultural identity, but neither embodies it strongly enough or takes it further than the historical moment, so that famine's violence is contained by the frame of the picturesque.

Whereas plays discussed so far in this chapter vary in the extent to which performing bodies may subvert the restrictions of rural Irish life, Friel's *Dancing at Lughnasa* is the clearest example. Its popularity in Ireland, England, America and elsewhere suggests it is more accessible – and perhaps easier to repeat – than *The Great Hunger*, which is more consistently unconventional in its physicality. First performed in the Abbey in 1990, then in London's National Theatre, *Dancing at Lughnasa*, directed by Patrick Mason, toured between two spells in London's West End, winning an Olivier Award, and transferred to America on Broadway. Frank McGuinness' film adaptation has also been screened by British TV and is available on video. Dedicated 'In memory of those five Glenties women', his aunts whom

Friel often visited in Donegal, the play is set in Ballybeg, the small imaginary community that *Translations* visited at an earlier date. Events are framed throughout by the adult narrator Michael, whose stage persona as a child is unseen. Actors do not acknowledge the narrator's presence but speak to the child's invisible body. This 'split self' narrative opens gaps, not only between the imagined and the real, past and present, but also – in Nietzsche's sense – that between the Appolline, controlling force of verbal language and the mysterious and creative potential of the Dionysiac body (1956). Michael is remembering a particular summer, poised on the brink of change, when at seven years old he lived with his unmarried mother Chris, her sisters and his uncle Father Jack – a missionary returned from Africa – and when his absentee Welsh father visited before going off to the Spanish Civil War. Crucially, therefore, these slippery memories of women from childhood are filtered through the discourse of an adult male. Whereas plays previously discussed in this chapter exposed the restrictions of rural life, imposed by necessary physical rigour as well as the force of economics and the church, here the emotional limitations and sexual yearnings that the latter factors repress are explored. Superficially this play might – mistakenly – seem gentler than the others. Although the link with colonialism is more subtly drawn than in the famine dramas, it is effected through bodily ritual, which ultimately reveals its destructive results. The film encourages bland misreadings of the play, partly through using an actual boy in Michael's role, and by misplacing the eponymous dance in the event sequence without paying precise attention to nebentext.

The West End production followed Friel's set instructions about revealing the cottage kitchen in section, with an outside garden space and tree. Further, the backcloth hanging behind some straw/standing corn stalks gave an impression of a corn-filled landscape, indicating the harvest period associated with the festival of the pre-colonial Celtic God Lugh. Basic authentic kitchen properties, including turf fire, signified the household's relative poverty, and seemed tonally dull in comparison to the brighter colour outside. A radio, named Marconi after the label, intermittently provided music, both traditional and commercial, suggesting another, more distant world. At crucial moments the women break out of the economic and emotional poverty of their conventionally restricted lives, and are temporarily transformed by different styles of dance that reveal aspects of their personalities. Comic Maggie dances frantically like a dervish; Agnes, gracefully evocative of her unspoken love; simple Rose, clumsily; Chris, subversively in a priest's surplice; and Kate, the most repressed and responsible, with complex but much tighter movements, closest to traditional step-dance, body rigid, feet frantic. She only dances once she has let out a repressed 'Yaahh!!' Friel's detailed nebentext describes this first outbreak of dance as

> grotesque, a pattern out of character and at the same time ominous of some deep and true emotion [...] there is a sense of order being subverted, of the women consciously and crudely caricaturing themselves, indeed of near hysteria being produced.
>
> (1990, pp. 21–2)

Prevented by Kate's concern for propriety and Father Jack's priestly status from joining Lughnasa festivities, at home the sisters erupt into song and dance that has a romantic and sexual power even when there is no music. What Kate calls 'old pagan songs' not only prompt dances but provide a playfully ironic edge to the domestic situation – for example, 'Anything Goes! (ibid., p. 64). Unfortunately, everything they have will eventually be lost. Although in the theatre it is easy for the audience to be swept up emotionally into the energy of this first dance, it is crucial to remember that Friel emphasises its grotesque and hysterical quality. It is not a purely celebratory dance, as the misplaced position and treatment in the film implies. In the play it follows a sad passage of dialogue when the women remember past dances and lost sexual opportunities – and is followed by a passage in which Kate tries to reimpose some kind of self-censorship, which is disrupted by the arrival of Michael's elusive father, Gerry. The dance is redolent of the forces of repression against which they respond with exaggerated embodiment of their different identities – hence the embarrassment and slight defiance mentioned by Friel. As suggested in analysis of *The Great Hunger*, the relationship between hysteria and the carnivalesque body is indicative of liminal states, that is, borderline moments suspended between change. Thus it is exactly appropriate for the liminal status that Michael at the end of the play attributes to this summer, in which dancing was a wordless ritual, a way 'to be in touch with some otherness' (p. 71). As emigrants, Rose and Agnes become socio-economic victims of a different kind of Otherness.

Although the women's dancing bodies have a quality of ritual, which mediates between harsh reality and lost dreams, Father Jack's role has most significance for postcolonial readings, as epitomized in the almost identical opening and closing tableaux. In the first, Jack has a dazzling British Army chaplain's uniform, so resplendent 'he looks almost comic opera,' Gerry (the absent father) has a tricorn hat with plumage. These are soiled and damaged in the final tableau, where the tatty uniform seems too large for Jack, who now carries the bedraggled hat. The play gradually reveals Jack has been sent home in disgrace, presumably because his behaviour and beliefs had become 'contaminated' by the life of the Africans he had supposedly been sent to 'convert'. He has lost much of his vocabulary and sense of place, but reveals that there he had been called 'The Irish Outcast' (1990, p. 39), refused to co-operate with the English and was suspected of going native by the District Commissioner, who had ironically given him the plumed hat. At the end of Act 1, Jack strikes out a rhythmic beat using Kite pieces.

> *With his body slightly bent over, his eyes on the ground, his feet moving rhythmically. As he dances – shuffles, he mutters – sings, makes occasional sounds that are incomprehensible and almost inaudible.*
>
> (ibid., 42)

Where the strictures of religion and patriarchy have thwarted his sisters' response to the carnival call of Lughnasa, a Celtic pre-colonial myth, Jack has been so seduced by the ritual of Africa that he is drawn to Ryangan sacrifice, drink and dance:

> The Ryangans are a remarkable people; there is no distinction between the ritual and the secular in their culture. And of course their capacity for fun, for laughing, for practical jokes – they've such open hearts in some respects they're not unlike us.
>
> (p. 48)

These potential similarities imply that in Ballybeg Dionysiac forces are largely repressed, yet in Africa, despite colonial rule, they have retained a dynamic resilience. The sisters' anxiety about Jack's behaviour may be an unconscious recognition of their potentially parallel situation and of such transgression, which they both fear and desire. Eventually accompanied by the sacrifice of Rosie's rooster, Jack's Act 2 exchange of tricorn hat for Gerry's straw hat is structured like a formal dance. Pine suggests Friel has modelled these Ryangan rituals on anthropologist Victor Turner's account of the Ndembu. This exchange is a rite of transition associated with a period of ambiguity typical of Turner's liminal phase and

> a sort of social limbo which has few (though sometimes these are crucial) of either the preceding or subsequent profane social statuses or cultural states.
>
> (Turner, 1982, p. 24)

Symbolic exchanges are seen by Baudrillard to be linked to ways in which human beings attempt to come to terms with death.

> We continue to exchange with the dead [...] we trade with the dead in a kind of melancholy, while the primitives live with their dead under the auspices of ritual and feast.
>
> (reprint, 1995, pp. 134–5)

Both Jack's Ryangan-influenced exchange of hats and the offstage power of the Lughnasa Celtic festival may be linked with human attempts to hold at bay and mediate between our uncertainties about the relationship between the powers of life and death. After his escape from Glenties, the adult Michael's story of his dead aunts is an attempt to achieve Baudrillard's impossible exchange – no one is ever 'quits with the dead [...] we never manage to 'return' what we have taken for all this 'liberty'' (ibid.).

Unwilling to recognise the significance of the coloniser's hat Gerry subverts it through Chaplinesque capers. Jack's ambiguous relation to colonising forces – even more so than Townsend's in *The Hanging Gale* – has left him on the borders between 'natural'/'civilised' and pagan/Christian behaviour. It would take more than a straw hat to incorporate him into Ballybeg. His rejection by the church causes Kate to lose her teaching job; his death precipitates the eventual household breakdown which is reported early in the play, thus breaking linear narrative and investing later scenes with irony. Transgression of Christianity is also manifested through female desire: Chris has an illegitimate child, simple Rose takes a lover – both against stereotypical notions of virginity and holy motherhood. Michael remembers that as a child he saw the romantic dancing of his mother and father as a kind of marriage, but like his father's promise of a bicycle, it was never fulfilled in a meaningful way. Ultimately, only deserted Chris, Kate and Maggie remain to scratch a living at home. Agnes and Rose suffer exile, poverty and homelessness in England after the factory takes over their home knitting. The final tableau of all family members, significantly different from the first, has the mythic function of blurring the boundary between actuality/illusion through an evocation of memory, which celebrates body rather than language:

> Dancing as if language had surrendered to movement – as if this ritual, this wordless ceremony, was now the way to speak, to whisper private and sacred things, to be in touch with some otherness. Dancing as if language no longer existed because words were no longer necessary ...
>
> (1990, p. 71)

This slippery memory can be set against the bleakness of post-colonial economics, which has destroyed the family, and is present in the final tableau through the subverted signs of imperialism and the kite's grotesque grinning face.

The third category of plays explores ways in which the slipperiness of spoken language may be linked with the body, but in an urban setting which prompts different strategies from those provoked by the rural landscape. Friel's *Volunteers* was not wholeheartedly acclaimed on its Abbey premiere in 1975, receiving 'almost universal putdowns' ((ed.) Murray, *Diaries*, 1999, p. 61). Although nebentext indicates time and place as 'the present in Ireland, an archaeological site in the centre of the city' (1989, p. 9), perhaps the play's presentation of political prisoners had overtones of the Troubles in the North – then not always a comfortable topic in the Republic. Seamus Heaney criticised reviewers who saw the play as being about internment (Pine 1990, p. 120). The play's border-crossing ideology pre-figures aspects of the Field Day approach, since it centres on excavation as a means of analysing shifting notions of the meaning/s of cultural identity. Discussion here centres on Mick Gordon's production at London's Gate Theatre during 1998, when the context of the Good Friday Agreement – an important stage in the Peace Process – intensified the play's significance for an audience of Londoners including those with Irish connections. The Director's programme notes that people at home in Northern Ireland are 'involved in the process of disentangling personal histories from ideological ones'.

The set provided an excellent example of Gilbert and Tompkins' view (1996, p. 146), that 'space/place' has a performative role in the way it determined interpersonal relations within its purlieu. The small, rectangular, fringe venue had seating arranged round three sides of what was virtually a traverse space. An earthy lower level was the site of the dig, while the 'office' space at the end near the entry door was at the same level as the audience, with the potential for actors to move round above the dig area while sharing the same space as the seating. Therefore the claustrophobic effect was close to Artaud's idea of seating the audience in the centre of the action to intensify contact between them and the actors, thus facilitating ritualised effects (1981 reprint, p. 74). Thus the audience was operating symbolically both as 'walls' of the womb-like or prison-like crater, and as a framing device for the play that laid bare the construction of both theatre and histories. As Nikolakis (2001) suggests, the audience acted like a Foucauldian panopticon, not only surveying the actors-as-prisoners, but participating in the power relations which keep them in place. Whereas George, the site manager, was relatively static, standing literally higher than the prisoners who have volunteered to work on the dig. Keeney, the leader of this group, and to a lesser extent Pyne his side-kick, were much more mobile around the set. Behaving like a theatrical double act, not only were their bodies expressive against the repressive limits of the state, but Keeney's excessive use of language, including both postmodern intertextual references and aspects of traditional seannachie-like story-telling skills, embodied similar tensions. Keeney's whirling words allude, amongst other literary echoes, to the enigma of madness in Shakespeare's *Hamlet*. The notion of the world perceived as a prison – like Hamlet's Denmark – is given further layers of meaning, especially as Keeney at one point addresses a skull. Keeney seems to embody some aspects of the 'Trickster', who appears in some other postcolonial literatures as a mediating/shape-shifting magical figure (Gilbert & Tompkins, 1996). Not only do some of his rhythmic and rhyming speeches have an incantatory quality, but he similarly roleplays a number of different personae, echoing the academic discourse of Professor King and Des the student, as well as setting up Pyne in role as a teacher, 'Flora (Tits) O'Driscoll'. Keeney's multiple personae range from bank managers to a naive American tourist, and to the report-writer, who has to assess him. His ability to capture different discourses runs through his pretended site Guide, who says 'these men are not what they seem' (1989, p. 36), underlining notions of fluid identity and ways that society forms individuals. Each prisoner is differentiated not only by his past, but his attitudes to the archaeological enterprise, and his relationship with the supposed histories of the corpse, Leif.

A relatively simple linear narrative shows this 'rescue' excavation's last two days, before suspiciously premature closure enables hotel-builders to take over the site. It includes tensions amongst the prisoners themselves and with George; the escape and voluntary return of one prisoner, Smiler, who has been beaten, institutionalised and brutalised to the point of social inadequacy. Further, these political prisoners face anxieties because their confederates have interpreted their voluntary involvement in the dig as treason and might kill them after it finishes. The focal point of both dig and set, is the revelation of the skeleton of 'Leif', who has been reached through the five-month excavation, as Keeney says

> from early Viking down to late Georgian – in other words over a period of approximately a thousand years'. So what you have around you is encapsulated history, a tangible precis of the story of Irish Man.
>
> (1989, p. 36)

The problematics of writing history are encapsulated in conflicting potential life stories attributed to the (so-called) Leif. Keeney's words spoken in the Guide role, suggesting that the more facts are known about Leif, 'the more comprehensive our definition of him' (ibid., p. 36), are undermined by the variety of these fabricated life stories. Keeney's suggestion that discovery of ancestors is a voyage of self-discovery illustrates the slippery nature of cultural identity. His first version of Leif's life in a Viking post and wattle house is relatively restrained. Pyne's more mythic version links Leif to the discovery of America and an Indian wife, whose Otherness is held fatally responsible for Leif's brother's death (pp. 6–63). Keeney later relates Leif's life as a viola player, son of a bankrupt merchant, who carried messages for subversives – thus unkindly echoing aspects of the life of Knox, one of the prisoners (pp. 69–70). Keeney taunts Butt, with a version suggesting that one way or another Leif was an exploited individual who stood up for his rights – until Butt exclaims Leif might have been one of the best in the movement, as Keeney had once been (pp. 71–2). Keeney's last version, which accompanies his rather solemn reburial of Leif, treats him like a contemporary friend (p. 83). The prisoners' final, speedy and contradictory litany of qualities starts with the idea of Leif as harmless, going on to his putative dubious relations, finally suggesting he was a bad seed, whose death can bear all their bad luck. Thus Leif, as scapegoat, is both everyone and no one. The mystery of his corpse provokes multiple answers; history can only be fractured, it is read according to different reader's 'take' on it. As Kearney suggests:

> Postmodern myth invites a plurality of viewpoints. It encourages us to re-read tradition, not as a sacred and inviolable scripture but as a palimpsest of creative possibilities which can only be re-animated and realised in a radically pluralist culture.
>
> (1988, p. 280)

One archaeological find has been reassembled by George. Keeney's remark, 'Smiler's pieces all put together and making a handsome jug' (1989, p. 55) suggests this mended jug (green in the Gate production) is symbolic of the escaped Smiler's chance of restored sanity. When Smiler returns, incapable of independence, Keeney despairs. Eventually Butt, who has throughout shown a seemingly genuine interest in archaeology, deliberately smashes the jug when George warns him to keep away from Keeney. The fragmented jug encodes the problematics of re-making stories of the past, the divided state of the island of Ireland, the shattering of friendships and allegiances, and the breaking of those who are involved in political struggles. Significantly Keeney's favourite form is the limerick, in which endings are often implied but not said. He does not deliver his final limerick's last line, leaving the volunteers' fate in question. But his last words (from *Hamlet*) 'Goodnight, sweet prince', presumably addressed to Leif, are undercut as afterwards George kicks away the stones and wood, which Keeney had placed to protect the corpse. The interplay of language, its excess and humour, including postmodern techniques such as intertextuality, gives *Volunteers* a deconstructive quality despite its apparently realist narrative drive. This particular production's efficacy was heightened by the claustrophobic staging, which enabled Keeney's diatribes to be delivered closely and threateningly to the audience, and by the external political negotiations.

Similarly, although Daragh Carville's *Language Roulette* seems to be a realist play set mostly in an urban pub, the dangerous power of language is a central theme. First performed in May 1996 by the Tinderbox Theatre Company at the Old Museum Arts Centre in Belfast, and revived in early 1997, it toured Ireland, then visited the Bush Theatre, London and the Traverse, Edinburgh. Winning a 1996 Stewart Parker Award, it was nominated as Best New Play in Barclay's TMA Awards, and as Best Drama Production, in Belfast City Council Arts Awards. The play's events occur in late 1994, after the Northern Ireland ceasefire. Although the nebentext indicates the main set is a Belfast pub, in opening scenes it doubles as the untidy flat of Colm, who is trying to tidy up while Ollie and Sarah his flat mates, watch TV. The apparent realism of this contemporary urban context is disrupted firstly by ironic music between scenes, often linked to a slightly uncanny repetition of the end of a previous scene at the start of the next. Secondly, the constant intertextual references to films add a postmodern deconstructive quality. These, like references to Joseph's European travel, suggest an environment beyond the postcolonial context – though he does acknowledge the latter, in a toast to '[...] the ceasefire. That's the most important thing' ((ed.) Fairleigh, 1998, p. 91). Peace, combined with a lack of helicopters, feels bizarre to these young people, who have always lived in the shadow of the Troubles.

Narrative involves Colm's plan for an evening's drinking in the pub with Joseph, returned from some years abroad, and another friend Tim. Not only do flatmates Ollie and Sarah insist on joining this reunion, but Tim secretly invites Colm's wife, Anna, from whom he is seeking an annulment, to join them later. Excessive drinking includes a range of games that facilitate power play among the group, becoming increasingly menacing as they remember past events rather differently, particularly the reason for Joseph's departure. Ollie and later Joseph collapse as a result of drugs offered by Tim – because, as is later revealed, he had in the past acted as a smoke-screen to conceal Anna's adulterous affair with Joseph and her consequent aborted pregnancy. The unclosed ending provides two repeated images of Tim, who had previously held a gun at Colm's head to force him to ask Anna for the truth, now with the gun at his side, and Joseph slumped.

Language Roulette, the play's title, is also the name of a game played abroad by Joseph:

> You go into a public place where there are crowds, like in the metro, and you choose someone you don't like the look of, and then you abuse them.
>
> (1998, pp. 93–4)

If the victim actually understands the language spoken, there is a chance that he/she would react violently. The technique is to say unpleasant things charmingly. This game highlights the slipperiness of language and its ability to hide the truth, which the play carries further. Joseph has already admitted he feels the way people talk 'at home' now feels alien:

> Well, it's just strange being back here when there's really only one language and we understand each other. It gives you an advantage when you have another language to hide behind. You can't do that here.
>
> (p. 93)

Tim's first game, in which the rules have to be guessed, involves each person speaking about whoever is sitting left of themselves, thus destabilising identities. This strategy gradually winds up tension as jokey verbal dexterity become more frenetic. Ultimately, the game 'Truth or Dare' also becomes a means of forcing repressed information dangerously to the surface under the influence of tequila slammers. Ollie's hospitalisation and Joseph's deterioration after Tim's drugs and oceans of drink lead to the evening's violent end: it is clear that despite the nominally shared language, the group of friends have not really understood each other at all. Thus both the nature of identity and communication are problematised.

In contrast, Walsh's *Disco Pigs* shows a friendship between two urban adolescents, bonded by their use of a private language. Produced by Walsh as Artistic Director of Corcadora, first seen in Cork, this play won Best Fringe Award at the 1996 Dublin Theatre Festival, transferred to the Traverse Theatre during the 1997 Edinburgh Festival, and then London's Bush Theatre. Walsh won the Stewart Parker Award (1996) and the George Devine Award for *Disco Pigs* and *Sucking Dublin* in 1997. At the Bush, a small venue, the audience, seated on two sides of the performance space, was bombarded by flashing lights and deafening disco music through much of the play. Spoken language mixed the local Cork accent with idiosyncratic sounds made by the male, Pig (Cillian Murphy), and the female, Runt (Eileen Walsh), but intensely physical performance made kinesics and proxemics more revealing of emotions and events – especially as events and interactions with others were not seen but conveyed by the pair. The narrative line, eroded by flashback/asides to the past, centres on the joint seventeenth birthday of the couple – friends virtually since their birth in 'Pork City' hospital. Their celebratory odyssey includes fast food, visits to Cork's pubs and nightclubs, a taxi ride to the sea, and the following day's visit to a Provo pub and more upmarket Palace Disco. Typical anarchic activities include a fight provoked by their deliberately staged jealousy scenario. On the second night, Pig, singing at the pub's karaoke, fails to prevent Runt being punched by a woman she has antagonised. Then, because she allows a man to kiss her hand at the Palace, he furiously beats him up, despite Runt's pleas. Finally Runt gradually realises she wants to be independent of Pig, and her speech gradually transmutes from Pigtalk to calmer, normal language:

> I look ... at the ducks ... as they swim in the morning sun ... in the great big ... watery-shite that is the river Lee ... Where to?
>
> (1997, p. 188)

as the lights go down separately on Pig, alone and distant, and then on Runt.

Certainly at the Bush, enhanced by the actors' close proximity, sweat and breathing contributed to the sense of the abject body conjured up by sounds uttered. Miming births to the sound of a heavy heartbeat started the show, and the sense of bodily presence developed from pig-grunts while eating amid speech, to scatterings of scatalogical words and a heavy use of plosives throughout. For example:

> Inta da skull like ka lawn mower it mix me and Runt all aboud! 2 fishys a swillin it back a swillin it back a swillin it back ... down da belly an oud da spout!
>
> (1997, p. 166)

Pig's explicit speech about sex with Runt emphasised the bodily nature and animal energy of youth, living within a restrictive and manufactured leisure environment. A still moment on Crosshaven sea-front contrasts with references to the 'bumhole of Cork city' and the violence of the fights. The satirical approach to the Provo bar – with music restricted to 'Danny Boy' and Provo songs – suggests that for these young people, the whole question of identity and postcolonial status is irrelevant. As Pig says:

> Ere, shouldn't ya be out plantin bombs an beaten up ol ladies yer fookin weirdos?
>
> (p. 180)

It seems therefore, that among some younger, urban-based Irish writers, there is an attempt to celebrate and experiment with the richness, flexibility and slipperiness of spoken language, either in itself or in a dynamic relationship with kinesics and proxemics. This type of approach, likely to appeal to new audiences, not only moves away from traditional tropes of landscape, but also from past issues, exploring characters who are now living

> in a not very generous society, North or South [...] severed from traditional loyalties and values, all would seem to be adrift in a new Ireland.
>
> ((ed.) Farleigh, 1998, p. xiv)

In contrast with this kind of focus on the new, which is discussed further in later chapters, the next chapter traces other ways in which fragmentation of traditional dramatic form and style has evolved to cope with representing the relationship between memory and politics.

3 Politics, Memory and the Fractured Form

Whereas the previous chapter explored the role of landscape and language, emphasising the performing body as a site of the inscription of power, this chapter is mostly concerned with the subversive function of dramatic form, and its relationship to the role of history and politics in the construction of identity. Catherine Belsey's analysis of ways in which so-called 'classic realism' is not value-free, but encoded with dominant ideology through both form and content, can be usefully related to the processes of postcolonial drama. Belsey points out differences between the classic realist and the interrogative text. She claims the former is characterised by 'illusionism, narrative leading to closure and a hierarchy of discourses' – it also tends to present a unified notion of human subjectivity, attempting to draw the reader into empathetic identification with dominant values thus voiced (1980, p. 70, pp. 90–4). The latter – and Belsey puts forward Brecht as a creator of the typical interrogative text – draws upon devices that, from time to time, distance the reader through exposure of the creative processes, thus challenging 'realism' not only through a deliberate lack of closure but through representation of competing and contradictory discourses. These challenge the reader to interrogate events and ideologies. Events do not necessarily follow a linear pattern, and may also merge elements from different time periods. Significantly the presentation of human subjectivity is represented as 'split' on the lines of Jacques Lacan's theories about language acquisition; revealing the difference between ' 'I' who speaks and the 'I' who is represented in the discourse' (Belsey 1980, p. 87). Taking this position into analysis of postcolonial discourses suggests that important dramatic strategies would be those that disrupt realism, and thus manifest fragmented form and subjectivity. Such texts therefore interrogate the colonising stamp of authority embedded in previous and dominant versions of history and politics. Following Bhabha's perspectives upon the role of split subjectivity in the postcolonial context (1984, 1990), Gilbert and Tompkins suggest that fragmented subjectivity not only reflects the 'many and often competing elements that define post-colonial identity' but that this split subjectivity may be viewed 'as potentially enabling rather then disempowering' (1996, pp. 231/2). Indeed, in one of their very few references to Ireland, they cite the role of memory and the fragmented self in Christina Reid's *The Belle of Belfast City* (1989), in which adult actors play characters in childhood and maturity, as indicating how replaying of subject positions 'serves to highlight the body's distinctions from stereotyped constructions of 'Irishness' (1996, ibid.).

Framing this chapter, Murphy's *The Patriot Game* (1991) and Kilroy's *Talbot's Box* (1977) introduce different types of challenge to realism in a post-colonial context, evident in contemporary Irish drama. Both are suggestive of the approach of European practitioners. The former, claimed by the author as documentary drama, is very much post-Brechtian in style, whereas the latter – despite deploying some Brechtian techniques – embodies more post-Artaudian, carnivalesque physical performance, which foregrounds space as a culturally specific site of transformation for postcolonial identity. This chapter focuses upon the relationship between private memory and public events, with some reference to internal and external forms of exile, including the significance of the revenant. The plays discussed use different strategies and are grouped under three headings: the problematics of writing history and politics, the intertwining of domestic memory and history, and the different deployment of ritual and carnival strategies to express political histories with mythic overtones.

The Patriot Game deconstructs what Murphy nevertheless hails as the 'birth of the Irish nation' (1992, p. xviii), the Easter Rising of 1916, which had just had its 75th anniversary. Despite his belief that the powers of nationalism and internationalism are not contradictory, his dramatic strategies, including the ironic title, effectively reveal the complexity of different ideological positions held then and now. Originally produced at the Abbey with thirteen actors doubling the many roles, Murphy suggests that a later production, using nine performers, was even more effective. Doubling is not merely a useful economic feature but demonstrates those socio-economic factors underlying roles to which individuals are assigned in life – and thus, as this Brechtian mechanism suggests, may be open to revolutionary change. Using actors for multiple characters encourages

> [...] fluid action and role changes in order to emphasise the performativity of the body and thus to frustrate viewers' desire for a fixed and unitary subject.
>
> (Gilbert & Tompkins, 1996, p. 234)

A female narrator, an alienating Epic convention, has throughout an ambivalent – almost disinterested – attitude to events she describes, is largely hostile to nationalism (1992, p. 129), and dances at the prospect of the British defeating the rebels as 'her victory' (p. 146). Near the end, after hearing the martyrs' final speeches, she is suddenly moved to cry 'Up the Republic!' so that the actor playing Pearse says to her ' [...] and you were only playing the narrator!' (1992, p. 149), a revealing slippage in the dramatic discourse. Conflicting viewpoints held during the Rising are shown through different recruiting aims of The Volunteers, the IRB and the Irish Citizen Army, as well as those MPs who supported sending soldiers for Britain's Great War effort. Disagreement, among both revolutionaries and the British, triggers muddle that prevented more widespread support of the rebels. In the Brechtian manner, source documents – key speeches and Pearse's poem for his mother – are used with songs, and include Gaelic. The nebentext indicates that actors playing a role may demonstrate an ironic attitude to his/her lines. For example, the actor playing Pearse's Mother is 'free in interpretation to question the sentiments' of the quoted poem (p. 115). At various points ritualised physical movements, like those linked with sound effects such as a drum, embody the advance of death, and stylised movement suggests fighting. The unclosed ending leaves the audience questioning not only the complexities of the historical situation and its heroic implications, but the problematics of representing it.

Talbot's Box, though representing the history of another man venerated in Ireland as a saintly individual, used performance techniques which, while also metatheatrical, have more Artaudian qualities and anticipate more recent physical theatre developments in Ireland. First directed by Patrick Mason at the Peacock in 1977, it is discussed in detail by Roche (1994, pp. 201–5). The nebentext makes the performance style clear. Set within the stage space, a huge box containing both all the actors and properties needed is the site within which Matthew Talbot at first lies in his coffin. 'The effect should be that of a primitive enclosed space, part prison, part sanctuary, part acting space' (1977, reprint 1997, p. 9). From this embodiment of darkness, dazzling lights emerge at certain moments, as do sounds of agony. Doubling is also a dramatic feature, since apart from Talbot the other two male and two female actors play a range of roles, including cross-gender slippage, and even a horse. Talbot, an unskilled Dublin worker, lived from 1856 to 1925. Born in severe poverty, at twenty-eight he abandoned his alcoholism through the Catholic temperance movement. Secretly he endured hidden physical penances, such as binding with chains, while also fasting and following strict routines of frequent prayer. Once these details were fully realised after his death, he became a cult figure, while moves towards his canonisation began in 1931. Kilroy's original aim was to write

> about the mystic and the essentially irreducible division between such extreme individualism, and the claim of relationship, of community, society
>
> (1997, p. 6),

but feels the play became essentially one about the cost and courage of being so alone. Although Roche notes Brechtian features, the performance style, with its emphasis on Talbot's tortured body, seems more akin to Artaud's ideas for the 'Theatre of Cruelty', or even Grotowski's notion of the actor's body as a sacrificial site (1968, 1986 reprint, p. 33ff). Further, dream-like elements – startling transformations such as a priest-figure's decline into a grotesque hag, contrasts between light and dark, echoing use of non verbal as well as verbal sounds – although with end-on if unusual staging, are also suggestive of Artaud's approach. He claimed, 'Violent concentrated action is like lyricism; it calls forth supernatural imagery, a bloodshed of images' (Artaud, 1970, 1981 reprint, p. 62). Strikingly vivid near the beginning is Talbot's return from the dead, succeeded by a dry lecture on the spectacle of his shrivelled body (1997, pp. 17–18). Actors' movements around the performance space, both inside and outside the box, symbolise various institutional and ideological pressures that attempt to contain/restrict Talbot's eccentric subjectivity. Throughout, words and actions are in tension with each other, mirroring the opposition between Talbot's individual, mystic vision and the Great Lockout of the Dublin workers in 1913, through which he lived – and worked – hence the taunt of 'Scab'. Thus embodying Talbot's retrospective dreaming of his life in flashback, the play reworks and questions that history, including the conflicting roles of the church and socialism within the colonial context prior to the Easter Rising. Fragmentation of form is ideal for challenging these ideologies, which can be linked to poverty and exploitation. Talbot, a kind of internal exile, is finally enclosed in the privacy of his box, while from the outer darkness, the other actors look 'in through cracks in the walls from which bright light comes which illuminates their faces' (1997, p. 63).

Kilroy has also used doubling to great effect in his more Brechtian play *Double Cross* (1986), a Field Day production in which Stephen Rea played both Brendan Bracken and William Joyce – men who recreated their own identities during the Second World War, finding their Irish identities intolerable. There is no indication these men met or knew each other, but Kilroy is interested in parallel aspects of their 'acted' lives, and their relationship with Germany. This play also explores the paradox of performing cultural identity.

Where *The Patriot Game* and *Talbot's Box* eschew realism, deploying predominantly either post-Brechtian or post-Artaudian strategies in their reworkings of history, which foreground public or private implications respectively, Reid's *Joyriders* (1986) and McGuinness' *Observe the Sons of Ulster Marching Towards the Somme* (1985) both show other ways of working against realist dramatic structures. The former's representation of Belfast youth is framed by a visit to a performance of O'Casey's *Shadow of a Gunman* (1922/3), while the latter draws upon an Ulster soldier's memories of fighting against Germany in the Great War. Both plays operate on the basis of a contrast between past and present.

Reid is a writer from Northern Ireland, but *Joyriders*, commissioned by Paines Plough, was premiered in the London's Tricycle in 1986, prior to a nationwide tour. Stephen Rea (as Devoren the supposed gunman), John Thaw (his friend Seamus) and Sheila Hancock (Mrs. Grigson) recorded 'Voices Off' from the performance of O'Casey's play. Songs used as a Brechtian strategy were created and performed with Clair Chapman by residents of Divis flats in Belfast, once described as the worst housing development in Western Europe. Events happen mostly where Youth Training Programme operates, the former Lagan Linen Mill. 'The mixture of Protestant/Catholic graffitti is due mainly to jokers [...]' (1997, p. 100), but Reid reminds readers a small number of Protestants are among the largely Catholic members of this course for youths (male and female) on probation or with suspended sentences for various petty crimes, including joyriding. Opening with the young people watching the end of *Shadow of a Gunman*, with Training Scheme Leader Kate, the nebentext indicates their varying responses (1997, p. 105). These reactions suggest difference in character and attitudes to the Troubles. Significantly, *Shadow of a Gunman* ends with Minnie, who took charge of a bag of explosives to protect the supposed Republican gunman whom she admired, being shot dead 'for love'.

Firmly set within the actualities of impoverished working class life, threats of violence from patrolling British soldiers, their check-points, and activists from either side of the religious divide, the framing context of O'Casey's play prefigures the outcome in the present. Combined with interspersed songs, which emphasise the Divis community's desperate state and their pent-up wasted energy, these two formal strategies move the play away from classic realism. They provoke questions about the right of young people not only to dream, but to achieve some kind of happiness beyond restrictions imposed by socio-economic forces and political difference. Despite representing different ideological positions, characters are more rounded than in strictly Epic theatre, and there is no doubling. Tommy, most politically aware, is also an opportunist thief. Arthur's steel-plated skull earns him both compensation for a soldier's 'accident' and eventually success as a chef through the Scheme. Cynical, tough Sandra prefers car mechanics to knitting, but hard-working Maureen, her mother shattered emotionally by a plastic bullet, hopes for a better future, although secretly pregnant by a foreign student. She has sole responsibility for Johnny, her glue-sniffing, joyriding twelve-year-old younger brother. Kate and her offstage friend, Claudie the social worker, are typically well-meaning middle-class women, once radical, now struggling in a reformist way to help disaffected young people, against the odds and official bureaucracy. Comic aspects are outweighed by the tragic penultimate scene, the Centre's 'Open Day', when Maureen – hoping to appeal to her lover in a new shoplifted outfit – is shot dead while attempting to intervene between soldiers' bullets and Johnny. Conflicting discourses of romanticism and realism are expressed in Sandra's angry cries as she shakes Maureen's corpse:

> This is what it's like. Do you hear me? It's not romantic and it's not lovely like in stupid frigging plays.
>
> (1997, p. 170)

The parallel with *Shadow of a Gunman* suggests bitter continuity into the present of a post-colonial struggle, which *Joyriders* shows is also connected to poverty of circumstances in 1980's Ulster, that not only denies viable life choices but is equally deathly.

Sandra's despairing cry and the elder Pyper's opening speech in *Observe the Sons of Ulster* reveal certain problems in writing with some degree of accuracy and fairness about history and politics, especially when conflicts in the present are rooted in different perceptions of the past. Pyper denies that he remembers anything from his past during the First World War:

> There is nothing to tell you. I am not a military historian [...] There are sufficient records, consult them. You are the creator, invent such details as serve your purpose best. Those willing to talk to you [...] to remember for your sake [...] they invent as freely as they wish.
>
> (1986, p. 9)

Where *Joyriders* borrows elements from Brechtian interrogative textual practices, while maintaining aspects of realism, *Observe the Sons of Ulster's* form is more radical in the way it destabilises notions of identity and the nature of historical memory. Directed by Patrick Mason at the Peacock, then the Abbey mainstage in 1985, it transferred to London's Hampstead Theatre in 1986, winning many awards. Revived in 1994 – as Murray comments (1997, p. 206) – at the time of the UVF ceasefire, it is about eight volunteers amongst the 6,000 Ulstermen of the 36th Ulster Division who were killed during the battle of the Somme on 16 July 1916, which coincidentally was the anniversary of the battle of the Boyne in 1690. McGuinness himself is a Catholic from Donegal, so in writing about this blood sacrifice made by Ulstermen, he is in some ways questioning claims made by Catholic Nationalists to be solely representative of 'Irish' heroism – as in Pyper's claim:

> His (Cuchullian's) blood is our inheritance. Not theirs. Sinn Fein? Ourselves alone. It is we, the Protestant people who have always stood alone.
>
> (1986, p. 10)

In interview with Kevin Jackson (*The Independent* 27 September 1989) McGuinness has claimed that he came to challenge 'his own bigotry' during the creative process (Roche, 1994, pp. 265–78).

The most significant dramatic strategies are firstly McGuinness' use of the 'split self', as two actors play the main role, Pyper, at different ages, and secondly the deployment of theatrical space as a site for transformation. Main events are framed by Elder Pyper's opening speech, and by final speeches when both Elder and Younger Pyper reflect upon and celebrate the Ulster for which the other seven men died at the Somme. Linear time sequence is broken as the ghosts of Pyper's dead friends emerge on stage, and growing relationships between the men are shown in the period leading up to battle. The importance of the revenant to the process of signification is discussed further in relation to *Mutabilitie*, later in this chapter. However, it can be seen as a return of the repressed in the Freudian sense of the uncanny (1985, Vol. 14 pp. 335–76), like the visions that return to victims of trauma such as shell-shock. Here, the reappearance of the ghosts and battlefield horrors, which Pyper has tried to deny, is linked to his guilt about his survival. He invites them to dance with him in the Temple of the Lord, a dance that is repeated at the play's end. Identity is explored throughout in a way that transcends questions of religious or political alliance: essentially the play is about growth and transformative change. The audience is invited to observe complex identities within what is often seen as a narrow-minded, embattled community. During 'Initiation', the second section, Pyper, whose own identity is hard to pin down, plays a deliberately mischievous role, somewhat similar to the 'Trickster' figure who challenges western traditions in much postcolonial literature. Such a figure 'evades and disrupts all conventional categories', may 'break down binary figuration of gender' and may embody 'subversive (and regenerative) power' (Gilbert & Tompkins, 1996, p. 235). Pyper's whirling words are not unlike Keeney's strategies in Friel's *Volunteers*, but his harrying of his fellow soldiers destabilises their sense of their own motivations and identities and their puzzlement about his.

The possibility of development in their self-perception is shown not only in the more literal context of barrack interchanges during which the men begin to form pair friendships, but is highlighted in especial moments that happen in liminal spaces. In 'Pairing', the pairs meet during leave from the front, in places in which their traditional values, and their different attitudes to their imminent return to war, are challenged and their identities transformed. These are real places, yet potentially heterotopic sites in which

> all the other real sites that can be found in a culture are simultaneously represented, contested and inverted. Places of this kind are outside of all places, even though it may be possible to indicate their location in reality.
>
> (Foucault, 1986, p. 24)

Spread across the stage, but lighted at different moments for their first couple dialogues, when seen again the nebentext indicates their simultaneous polyphonic conversations, each of which is partly heard by the audience as the lights go up on all areas (1986, pp. 52–61). While looking at ancient carvings on Boa Island in Lough Erne, Craig and Pyper's possible homosexual relationship is cemented; in a church, Roulston, the doubting Protestant ex-clergyman, and Crawford, the 'half Fenian' possible Catholic, grow closer. Moore, shattered by his military experience, is encouraged by Millen to cross a bridge over an abyss, until eventually Millen too realises their reciprocal need for friendship and support. McIlwaine and Anderson, belatedly celebrate the Orangemen's commemoration marches of 12th July, with their traditional lambeg drum. Part Four, 'Bonding', shows the unity of these diverse identities in collective activities: football, a mock Battle of Scarva – in which William of Orange triumphs – and singing a hymn. These activities culminate in a final ritual exchange of Orange sashes, underlining the way in which all the men go into battle, celebrating their love of Ulster. Since these men are fighting for the British army against the Germans, it might seem paradoxical to consider the play as a postcolonial drama. Crucially, McGuinness sympathetically represents a historical moment in which the complexities of identity resonate with the particular Irish political context, but also those beyond it. His approach mixes postmodern dramatic strategies with history-based concerns that are surely postcolonial.

Two productions that call for a more detailed examination of the way in which fragmented form and subjectivity heighten memory's role in the problematics of representing history are Barry's *The Steward of Christendom* (1995) and McGuinness' *Mutabilitie* (1998). Both can be considered in the light of their dramatic use of split subjectivities and heterotopic space, which can in the latter instance be more closely linked to ideas of Homi Bhabha and Baudrillard. Both are thus clearly postcolonial dramas – and examples of Kearney's mediational modernism. With five dramatic works already performed in Ireland and also America, *The Steward of Christendom* brought Barry to a wider audience. Originally produced by Out of Joint in conjunction with London's Royal Court, directed by Max Stafford Clark, it expresses the relationship between personal memory and history most poignantly, winning rave reviews for both its author and actor Donal McCann as Thomas Dunne. Awards won included Writers' Guild 'Best Fringe Play 1995' and Critics' Circle 'Best Play 1996'. Nominated for an Olivier Awards and BBC 'Best Play 1995', Barry also won the Ireland Funds Literary Award and the Christopher Ewart-Biggs Memorial Prize. The play's brilliant exploration of the slippages between memory, history and identity support Barry's view

> I believe more in how the mind remembers things than I do in that masculine surety, written history.
>
> (Interview)

It is drawn from the situation of Barry's great-grandfather, Thomas Dunne, last Catholic Head of the Dublin Metropolitan Police before the change of regime in 1992 – and thus responsible for Dublin Castle when it symbolised British rule. Dunne is represented as a split subjectivity which is directly linked to the colonial past, while the use of theatre space epitomises the merging of his memories of private and public events. Transferring from the Royal Court Upstairs studio to the mainstage, the set's depiction of the grimy walls of Baltinglass Asylum, County Wicklow, where in 1932 Dunne is now confined, became partly translucent as the borders between past and present dissolve. Traces of golden light suggested aspects of landscape evocative of his earlier and happier rural past, while the grey walls had an almost imperceptibly tiny trace of gold along the skirting and door frame. Through the walls of the simple room with its central bed, a stool and table downstage left, Dunne's dead son and his daughters appear across time, interspersed with Dunne's present-day interaction with the warders: kindly Mrs O'Dea, and more irascible Smith – otherwise 'Black Jim' – who enter through the door, to the right. Dunne's memories disrupt linear time, merging a variety of spaces, creating a borderline, heterotopic space which, like the asylum, is nowhere, but yet everywhere. As Foucault puts it, this 'placeless place', mirrorlike, enables Dunne to experience seeing himself

> [...] in an unreal space that opens up behind the surface; I am over there where I am not, a sort of shadow that gives my own visibility to myself.
>
> (1986, p. 24)

Although critics of Barry's work generally tend to emphasise verbal elements, remarking on poetic rhythms that may echo Synge or James Joyce, here, as in his other works, striking visual moments often created by lighting effects are indicated through the nebentext. Barry himself disputes the description of his plays as 'poetic', stressing they show how humans speak about ordinary but powerful experiences: 'Poetic things do not reside in language, but within us' (Interview). *The Steward of Christendom* has, like some of his other plays, liminal moments arising from a quality of waking dream/reverie, which epitomise Barry's belief that seeing and hearing are pre-literate – and that playwriting joins his 'expertise as a mature individual to unconscious experiences as a child ... as a human being' (Interview). In the late Donal McCann's outstanding performance, Dunne's body encapsulated the paradox of the colonised body from the moment he is first stripped. In dirty long-johns, suggestive of bodily accretions associated with rare washing, big-framed McCann's repetitive gestures – alternating between an almost foetal crouch and a quasi-military erect posture, sometimes moving hand to head to convey his confusions – were the embodiment of repressed guilt.

It emerges that because Dunne's beloved wife had died in giving birth to his favourite daughter, Dolly, he had been left to bring up his three girls and his son, Willie, who died young during the First World War, and to whom he had not been close. Dunne's position as Superintendent had damaged the lives of these three loved girls through limiting their social opportunities: plain Maud escaped through marriage to artist Matt Kirwin, Dolly escaped to America, but single, polio-crippled Annie looked after him. Dunne's pride in his loyalty to the British Crown and especially to Queen Victoria runs counter to the period's Nationalist spirit:

> Her mark was everywhere, Ireland, Africa, Canada, every blessed place. And men like me were there [...] to keep order in her kingdoms [...] Ireland was hers for eternity, order was everywhere, if we could but keep her example.
>
> (1995(a), p. 14)

Despite pride in his own controlling violence, he felt the charisma of Michael Collins at the moment when he had to hand Dublin Castle over to him;

> Most of the men in my division were for Collins. For an instant [...] I felt a shadow of that loyalty pass across my heart.
>
> (p. 50)

Thus his dislocated memories and repressed admiration often return to the assassination of Collins:

> I think that the last order I gave to the men was to be sure to salute Mr. Collins coffin as it went by.
>
> (p. 63)

This inner contradiction is echoed in his feelings of failure as a father, and regret that his grandchildren are now not allowed to see him. His memory recreates key scenes: the moment his daughters help him don uniform to relinquish the castle, or when Dolly shows him her travel ticket to Ohio. Past domestic interactions – with a police recruit, Willie singing, or faithful Annie's visits despite her father mistaking her for long absent favourite Dolly, are juxtaposed with present events.

Dunne's regrets about 'the things of my own doing and damned history' (p. 10) are throughout foregrounded by his body. In anguish he roars wordlessly, beats bed and table – or threatens imagined fellow countrymen coming to punish him for his years as a traitor, enforcing the colonizers' rules. These fears and his request that Annie strike him with his ceremonial sword had been the immediate cause of his confinement. The warder's beatings seem an ironic parody of his own imposition of law and order of which he had been so proud. Dunne's position is suggestive of what Kiberd defines as a kind of defeated, emasculated power,

> [...] a posture of provincial dependency as a policeman, or a bureaucrat or a petty official in an oppressive or despised colonial administration.
>
> ((ed.) Keneally, 1992, p. 132)

Refusing to be clothed in an asylum suit, Dunne still wishes for a trace of gold braid, which he felt had not been plentiful enough on his dress uniform – a wish that kindly Mrs O'Dea satisfies with yellow cotton tacking. The power of costume as role is later undercut when Smith, feared by Dunne as 'Black Jim', dressed as Gary Cooper for an offstage party, anoints Dunne's wounds with unaccustomed kindness. This costume device of a fantasy hero helps to distance and provide an ironic opposition when Smith reads aloud Willie's long-preserved letter from the war in which he had written,

> I wish to be a more dutiful son, because Papa, in the mire of this wasteland, you stand before my eyes as the finest man I know, and in my dreams you comfort me and keep my spirits lifted.
>
> (p. 57)

Final, touching moments show Willie, the revenant, dressed in his First World War uniform, muddy but with a tinge of gold, climbing on his dying father's bed, while Dunne reminisces about the one moment of mercy that he recalls, when his own brutal father desisted from beating him as a child, although he had run away to protect his pet dog who savaged sheep,

> And the dog's crime was never spoken of, but that he lived until he died. And I would call that the mercy of fathers, when the love that lies in them deeply like the glittering face of a well is betrayed by an emergency, and the child sees at last that it is loved, loved and needed and not to be lived without and greatly.
>
> (p. 65)

As Dunne finishes he falls into sleep, or perhaps death, as Willie's ghost lies close to him in fading light. In a sense, his exile from himself is over. Such a transcendent moment of forgiveness and reconciliation encapsulates Barry's belief that 'Triumphalism does not produce grace' (Interview), a positive perception relevant to any process of negotiation, especially in the post-colonial context.

The closing moments of McGuiness' *Mutabilitie* also link a child's presence with potential forgiveness, but whereas a kind of 'internal exile' is one of *The Steward of Christendom*'s themes, here the kinds of exile represented are both more literal and more complex. Foucault claims after experiencing split subjectivity through a heterotopic space, 'I begin again to direct my eyes towards myself and re-constitute myself where I am' (1986, p. 24). This moment of time lag, during which the gaze goes back and forth, is similar to that in which Bhabha suggests that the 'sign seizes power to elaborate [...] new and hybrid articulations and agencies' (1997, pp. 191–2). Rooted in analysis of what Bhabha calls 'the Third Space' (1994), he further claims the 'split' or destabilising of the subject has a similar transformative power. From this book's perspective, that Bhabha relates this liminal, in-between space to cultural hybridity is crucial to the use of performative space in postcolonial drama, because he claims:

> Contingent, borderline experience opens up *inbetween* colonised and colonised. This is the space of cultural and interpretive undecidability produced in the 'present' of the colonial moment.
>
> (1997, p. 206)

Mutabilitie, set in a historical period, embodies these qualities assigned to the heterotopian and Third Spaces, such that McGuinness' postmodern – or mediational modernist – theatre strategies manifest postcolonial potential. In particular, the ways that different kinds of exile destabilise and split the subjects of this drama, facilitate the possibility of transformative change. Deployment of theatrical space here also evokes the link between representation and death through the role of exile and the revenant. The mixture of history and myth triggered unfavourable reaction from critics who failed to recognise the play's strength lies in its web-like qualities rather than in narrative. Charles Spencer wrote,

> The hapless McGuinness appears to have been possessed by the spirit of that ranting wildman of the theatre Howard Barker [...] the play is over-written [...] windy rhetoric'.
>
> (*Daily Telegraph*, 24 November 1997)

As the title suggests, the play celebrates the slippery qualities of memory, myth and history. Staged at the Cottesloe studio of the National Theatre, London in 1997, and set in Ireland during 1598, it begins with the arrival of three Englishmen, one of whom falls into a river and turns out to be William Shakespeare, a gay, closet Catholic. The other two, actors, are captured by followers of a dispossessed Irish chieftain King Sweney, and his wife Maeve. Shakespeare is taken into the home of the English Protestant poet Edmund Spenser, author of the epic *The Faerie Queene* (1591–8), which includes the *Mutabilitie Cantos*. Shakespeare is mistaken by the Irish poet/prophetess The File as a foretold saviour/poet of the Irish, who are Catholics. She and her husband, Hugh, ostensibly servants of Spenser and his wife Elizabeth, are using the opportunity to work on the coloniser's weaknesses. Similarly Annas, daughter of King Sweney, seduces and betrays one of the two actors so that both are killed. Ultimately, the Irish chieftain's sons kill both parents at their request. Set alight – perhaps by himself – Spenser's castle burns down and he, his wife and family flee. The File helps Shakespeare to escape, and the dispossessed remaining Irish tribe adopt rather than kill Spenser's surviving son. This complex, rather fragmented narrative includes multiple verbal and visual discourses, such as intertextual references to history, myth, ancient classics, Irish and Elizabethan literature, songs both English and Irish, a metatheatrical interlude on the fall of Troy, a variety of sexual inter-relationships – even perhaps incest – as well as a melange of different linguistic references and registers.

Edmund Spenser (1552–1599) did indeed spend much of his life in Ireland as secretary to Lord Grey of Wilton, a ruthless executioner of Irish rebels, and was rewarded with Kilcolman castle and manor. His treatise *A View of the Present State of Ireland* (1596, publ. 1633, (ed.) Hadfield 1997) applauds savage colonial methods, pondering upon managing 'the stubborne natione of the Irish to bring them from their licentious delights of barbarisme unto the love of goodness and civilitie' (1997, pp. 20–1). Both treatise and Spenser's *Mutabilitie Cantos*, written after the castle fire prompted his return to England, permeate McGuinness' play. The opposition of two cultures who see each other as the epitome of 'Otherness' is heightened through unconnected mythic versions of Irish history. These are embodied in Sweney, a king from a saga originating from the battle of Moira (AD 637), and Maeve, the Connacht warrior Queen from the Epic *Táin Bó Cuailnge*. The portrait of Shakespeare is uncanonical – and history has no record of a visit to Ireland. Overlaying of these genres, times and spaces is appropriate to modern times: 'We are in the epoch of simultaneity, we are in an epoch of juxtaposition' (Foucault 1986, p. 22). The historical moment when the play is set marks the break between the concepts of stasis and infinite movement by deploying the late sixteenth and early seventeenth century idea of mutabilitie, which foregrounds the everchanging motion of the apparently static space. However, the twentieth century moment of reading this performance marks the break between notions of linear progression and those of relativity and fragmentation. As Foucault implies, for the twentieth century, termed the 'epoch of space', temporality is less significant (1996, p. 23).

Monica Fawley, often associated with the Irish National Theatre, designed a traverse setting that blurred myth and history, merging in Foucault's sense 'several sites that are sites that are in themselves incompatible' (1986, p. 25), so that space is thus representative of 'the form of relations between sites' (ibid., p. 23). These 'varous distributive operations that are spread out in space' were here associated with either coloniser or colonised. On one side of the traverse, Spener's castle Kilcoman was represented in a cardboard-style edifice, which deliberately gave the impression of a child's 'pop-up' book. A drawbridge, sometimes lowered to reveal the interior with a large copy of a portrait of Elizabeth 1/Gloriana upstage, at different times used patently two-dimensional schematic props and furniture, such as a desk with writing materials or crockery. Spread across the centre space was an uneven rocky terrain with quite deep water-filled hollows. At the other end, an earthy area including apparently living trees was the forest where the dispossessed Irish lived. Critics who dismissed this set as 'designed for a EuroDisney ride' (Roger Foss, *What's On,* 26 November 1997), or complained 'reminiscent of a computer game [...] close up the figures look too heavily costumed, like figures in a heritage museum' (Robert Butler, *The Independent on Sunday*. 23 November 1997), failed to recognise this kind of visual design, which deliberately drew attention to itself, foregrounded the slippery and changing nature of history and myth as constructed narratives. Spatially the set embodied binary oppositions: artifice/nature, English/Irish, coloniser/colonised. The four elements earth, air, fire and water were used symbolically as parts of ritual processes linked to transformation and death. Physical movements required by those who needed to struggle from one side to the other – such as the Irish servants Hugh and the File who were also spies – not only emphasised metaphorically the difficulties of making connections between cultures, but also their spatial separation showed how a combination of different existential and historical time

> even locally, does not slide up and down a temporal scale [...] but jumps back and forth across a game board that we conceptualise in terms of distance.
>
> (Jameson, 1991, pp. 372–3, qtd. Bhabha, 1997, p. 218).

The traverse facilitated a wide spread of actors, as at the end of Act 3, where a polyphony of voices set up through pairings in tableaux poses, visually and aurally demonstrated the heterogenous network of relationships/sites indicative of the post-colonial condition. Songs – pagan, Catholic and Protestant – were also sung in counterpoint from different parts of the stage.

Complex intertextuality operated at several levels, language registers veered from the poetic to contemporary and modern vernacular, often to humourous and deconstructive effect, or tinged with ironic significance. Echoes of Spenser's writings evoked his contradictory love/hate relationship with Ireland, in which he mythologises the Elizabethan colonial project. Spenser, Shakespeare and the other actors are in exile from England, the Irish King and Queen from their own Kingdom, Hugh and the File in apparent exile from their own people. These characters are also in exile from themselves, their identity is uncertain. Spenser, echoing Shakespeare's *Henry V*, asks ' What is my nation?'; Sweney asks, 'Who am I? I have forgotten'; Shakespeare laments, 'I can't remember who I am'.

References to theatre as the place of the revenant heighten the unsettling link between representation and death. The File commands, 'In this your theatre you will make our dead rise, William' (1997, p. 61). Edmund, haunted by his father's ghost, almost kills his own child. The File suggests to Spenser's wife, also named Elizabeth, that Munster, laid waste and 'replanted' with English is haunted by the voices of the dispossessed Irish. Spenser, supporting brutal repression in Munster, in both his treatise and this play describes the colonised as revenants:

> They looked anatomies of death, they spoke like ghosts crying out of their graves.
>
> (McGuinness 1997, p. 12 & pp. 78–9, Spenser (ed.) Hadfield, p. 101).

A corpse, like the double or the revenant is in Lacan's terms, a reminder of the initial absence of the mother's body, which his theory links with the moment when the child learns to speak in language, and thus to the splitting of the speaking subject (1977). Poised in the gap between social and psychic identity, corpse and revenant are thus reminders of the instability of identity, of mutability and change. As the File sings:

> I call on Death, most trusted friend
> To bring your exile to its end,
> Mankind, the sky, the rivered sea
> Sing of Mutabilitie.
>
> (1997, p. 43)

Bodily performance emphasised sexuality, charged with death and linked with an eroticism of the Other – a feature common to the colonial situation. Thus the liminal qualities of this heterotopic space could be seen to be synonymous with the gap or slit between signifier and signified, within which, according to Lacan (1979, p. 37) the 'rhythmic pulsation' of appearance/disappearance, continuity/discontinuity, death/desire oscillate. Bawdy talk embraces promiscuity, homosexuality, seduction, possible incest, and Elizabeth the First's sexual practices, in contrast to Spenser's abstract fantasies about the Virgin Queen. Nevertheless, the various exchanges of desiring bodies in the play are not powerful enough in themselves to generate transformation.

Baudrillard (1995, p. 126) has suggested twentieth-century Western capitalism exists on a sharp binary separation between life and death, and thus on the notion of discontinuity, whereas previously rituals of exchange kept communal identity alive. The seventeenth century was for him the crucial moment when

> The primitive thought of the double as continuity and exchange was lost, and the haunting double comes to the fore as the subject's discontinuity in death and madness.
>
> (ibid., p. 142)

Whereas Edmund's seemingly more modern Oedipal struggle suggests the need to shake off guilt and establish a discontinuity with the future, Maeve's demands that her children kill Sweney and herself can be seen as consonant with Baudrillard's concept of the continuity of ritual exchange as in earlier times. Their deaths have been given in a 'transmutation of the flesh into a symbolic relation, the transformation of the body in social exchange' (ibid., pp. 138–9). As in Spenser's *Mutabilitie Cantos* the challenge of mutability is eventually seen not as discontinuously destructive, but as part of a continuous natural cycle:

> Then over (things) Change does not rule and raigne
> But they raigne over Change and do themselves maintain.
>
> (Spenser (ed.) Yeats, undated, p. 179)

By seizing power to command their own death, the Irish challenge to the coloniser here demands an equal exchange, defying the system 'with a gift to which it cannot respond save by its own collapse and death' (Baudrillard, 1995, p. 37). It is the climax of a framework of exchanges: dispossession of the Irish balanced by the flight of Spenser's family from their burning castle, the murder of the two actors balanced by the offstage, ritual 'sacred' murder of the King and Queen. After a ritual in which the Irish relinquish war and embrace destitution, a reiteration of family love accompanied by a scene of bathing/purification occupies the central space. Such resistance to the 'ethics of accumulation' is the opposite of the way Baudrillard suggests the modern mind defends itself against death. The exchange scenario is completed by Irish fostering of the live, lost English child, as an exchange for the File's baby, maternally murdered during the wars. In the face of this 'immediate non-phantasmic actualisation of symbolic reciprocity [...] Everything is there, reversible and sacrificed' (ibid., p. 145). Fed on the milk of human kindness, the child is to be as it were a cultural hybrid 'fostered as our own' (McGuinness, 1997, p. 93). Thus, in tune with Bhabha's terms, the mutability of the Third (and heterotopic) space has enabled

> [...] through a structure of splitting and displacement [...] the architecture of the new historical subject (to emerge) at the limits of representation itself.
>
> (1997, p. 217)

McGuinness closes the play with the transformation of the child as a sign of a new and more complex identity: a herald of postcolonial subjectivity/subjectivities. Such identities are in tune with Kearney's postnationalist, hybrid

> 'imagined' community which can be reimagined again in alternative versions.
>
> (1997, p.188)

In the second group cited in this chapter's second paratraph, two plays with contrasting forms, which also use the idea of the revenant as a strategy for exploring the link between past and present, but use a domestic context, are Stewart Parker's *Pentecost* (1987) and

Dermot Bolger's *April Bright* (1995). Set in Belfast and Dublin respectively, both plays explore a house invaded by the past, so that linear narrative is fragmented. Parker – to whose striking talent the prestigious Award has been dedicated – during his lifetime (1941–1988) created nine theatre plays, nine radio dramas as well as TV dramas. In the posthumously published *Four Plays for Ireland* (1989), Parker, who considers *Pentecost* to be an example of 'heightened realism', suggested,

> Plays and ghosts have a lot in common. The energy which flows from some intense moment of conflict in a particular time and place seems to activate them both.
>
> (1989, p. 9)

Set in Belfast his home town, *Pentecost* is the last of a history trilogy, embodying Parker's sense both of the presence of ancestral voices, and the pressing need for reconciliation. Parker's 1986 John Malone Memorial Lecture at Queen's University, Belfast, expresses his artistic aim:

> The politicians, visionless almost to a man, are withdrawing into their sectarian stockades. It falls to the artists to construct a working model of wholeness by means of which the society can begin to hold its head up in the world
>
> (qtd. Roche, 1994, p. 220)

Paradoxically, through representation of fragmentation, Parker stresses the possibility of an open future because as Marian suggests,

> We don't just owe it to ourselves, we owe it to our dead too, the innocent dead.
>
> (Parker, 1989, p. 208)

Directed by Patrick Mason for Field Day, with a cast including Stephen Rea and Eileen Pollock in 1987, it went from Derry to Dublin Theatre Festival, then with a different cast to London's Tricycle in 1989. There, rather cramped stage space conveyed the cluttered almost suffocating atmosphere of the old house, as in the nebentext:

> Everything is real except the proportions. The rooms are narrow, but the walls climb up and disappear into the shadows above the stage.
>
> (Parker, 1989, p. 147)

Marian and Lennie, a musician, are both Catholics, but their marriage is virtually dead. He has inherited the house, via a landlord relative, after Lily, the last tenant's recent death. An old Protestant widow, she had kept the house as it was when her husband died in 1959. Marian wants to buy the house from Lennie as a kind of emotional refuge, originally intending to negotiate with the National Trust so that it could become a record of 'a whole way of life, a whole culture [...] a greater community of experience' more valid than that of an aristocratic home. The house also becomes a place of refuge for her old friend Ruth, a Protestant regularly beaten up by her policeman husband. Lennie's friend Peter, another Protestant, returns from Birmingham to experience historic events at first hand, and is accused of holding the British attitude, 'A plague on both your houses' (Parker 1989, p. 185). The house itself is situated in a borderline position, described by Marian as 'eloquent with the history of this city' (ibid., p. 165). Lennie points out its dangerous location,

> It's the last house on the road left inhabited! – The very road itself is scheduled to vanish off the map, it's the middle of a redevelopment zone, not to mention the minor detail it's slap bang in the firing line, the Prods are all up in that estate, the Taigs are right in front of us [...]
>
> (ibid., p. 154)

Personal relationships are set against important public events, around the Ulster Worker's Council Strike of 1974. Following Bloody Sunday, a period of Direct Rule from Westminster had been imposed in late March 1972. Although a power-sharing Executive was formed in December 1973 from the Northern Ireland political parties, it crumbled away, partly as Murray suggests (1997, p. 218), as a result of this Strike, thus causing resumption of Direct Rule. History's shadow, both public and personal, is embodied in Lily's ghost, whose unfinished cup of tea and knitting suggest a need for completion. Seen and heard only by Marian, she in some ways epitomises her 'other self', especially as it is gradually revealed that both of them have lost a child. Marian imagines memories of Lily's life on the basis of items found. This doubling of personal experience is echoed in the sexual passion that Lily had shared with her lover in the past, and that between Peter and Ruth in the present. The friends react angrily to a BBC broadcast of an inept speech by the British Prime Minister, Harold Wilson, and to sound and lighting effects, which convey offstage turbulence in the streets. The wound inflicted by Ruth's husband is echoed in damage inflicted on both Marian and Peter, who ironically displays his 'red hand ', a symbol of Ulster. Critics disagree about the efficacy of Biblical passages quoted towards the end. Nevertheless, the invocation of Pentecostal flames and visionary tongues operates as a ritual, drawing in the audience whether in Derry, Dublin or London – whose ideological positions may be as different as those of the reconciled characters on stage – to feel that some kind of resolution may be possible. The light of the future is let not only into the stage house but into the outside world; the opened window of the ending implies the possibility of reconciliation, but does not, as in classic realism, define it.

Although Bolger is also a poet and a novelist, the complex nebentext of *April Bright*, originally shown at the Peacock in 1995 before an Irish National tour, clearly demonstrates his skills in dramatic visualisation. Here a Dublin house is haunted not by past political tensions, but a socio-economic history associated with endemic poverty and consequent poor health. The play's critique of the socio-medical hierarchies of the 1940s is filtered through memory and contrasted with the present through a split subjectivity – the Caller. As in *Pentecost*, a present-day young couple is in the process of moving in, but the past is gradually revealed. A prolonged visit from a mysterious Caller – whom the audience may later deduce to be Rosie Bright – is a barren woman who had lived there as a child during the 1940s. Anna, pregnant but uneasy both about the level of commitment in her relationship with Sean and the possibility of another miscarriage, is sensitive to atmosphere. Gradually snatches of the Bright family's life are revealed, intermingling with present-day interactions of Sean and Anna, who cannot see them – although the Caller can, including her young other-self, Rosie. Set as a split-level house with the kitchen area downstage, and bedroom on a raked area further upstage, both levels have hidden entrances which enable revenant figures to appear 'as if ghosts from nowhere' (1997, p. 9). Performance timing needs to be tight yet fluid, as at various moments the living as it were almost feel the dead's presence, as they change places at table or even touch them. Poses of the 'living' sometimes echo those of the 'dead' across three time-frames, 1940, 1970 and the present.

Slowly through these not entirely linear 'remembered scenes', the audience learns that, despite her energetic, mischievous character, April, Rosie's sister, was diagnosed with fatal tubercolosis, not then uncommon in crowded housing. Eamon Bright, having already lost a young son from Rheumatic Fever, despairs. Despite sibling rivalry, Rosie and April had been happy together as children, especially in their bedroom, imagining their future partners and children. When the hospital sent the sick April to the Poorhouse due to non-payment of fees, without notifying the family first, Eamon converted the garden shed for April's isolation. Apparently families were often forced by financial pressures and shame associated with TB to adopt this strategy. Since April's death, the Caller has retained an antipathy to wooden huts, but also remembers trying to persuade her dying sister that life was still possible, by saying,

> I don't want this house April. May God strike me barren, may he dry up my womb
> before a child of mine runs up these stairs
>
> (1997, pp. 115–16)

Now she reassures Anna her expected baby will bring happiness back to the house. Thus prompted, Anna and Sean come to a new, more mature understanding of their relationship. The final double-image contrasts April in her bedroom, showing her imaginary but never realised baby the bright and beautiful world outside, with Anna downstairs also holding up her imagined, but due to be realised, baby to the light, 'Our precious child to come' (1997, p. 120). Again, this play celebrates future potential through hints rather than a prescribed outcome.

That private and historical events are inevitably interwoven can be seen in contrasting plays from the Republic and Ulster respectively, which explore the pull between family relations and ideologies. Barry's *Prayers of Sherkin* (1990) and Ann Devlin's *After Easter* (1995) both focus upon a female protagonist: in the former she leaves her family for ever; in the second, she returns home to confront her problematic identity. Directed by Caroline Fitzgerald for the Abbey, *Prayers of Sherkin* explores the past through Barry's great-grandmother, Fanny Hawke, who left the Protestant community on Sherkin island. 'The thought struck me that if she hadn't crossed that narrow stretch of water, I wouldn't exist myself [...]' (1997, Interview, Mick Moroney, p. 4). From scraps of family history Barry created 'the very shadow of her true life, a piece of old thread'; which yet rang true with those who remembered her. In 1890 the Sherkin community, founded by Matt Purdy with three families from Manchester one hundred years previously, has dwindled to five members of the Hawke family, candle-makers for the mainland. Only Fanny and her brother Jesse are of marriageable age. Events focus on her decision to marry Patrick Kirwin, a half-Jewish Catholic lithographer from Cork, who falls in love with her when she visits the mainland. Marrying out of the community entails leaving forever. Small details, such as the aunt's weakness for ribbons or Jesse's fascination for telescopes, and pervasive imagery of nature, sea and light convey the simplicity of island life and its gentle humour. Fintan O'Toole commented on the original production's luminously beautiful quality, but English critics were more divided about Peter Hall's 1995 London Old Vic production, suggesting its 'parochial quaintness and narrative slenderness' (Nick Curtis, *Evening Standard*, 20 May 1997) lacked the 'conflict that added depth and vigour' to *The Steward of Christendom* (Maggie Gee, *Times Literary Supplement*, 6 June 1997). These misreadings failed to recognise that the play's strength lies in the tale's economy and sparse Shaker-like setting which through fluid lighting and poetic, lyrical qualities, suggested moments of transcendence. Some of these are also triggered by the presence of a revenant, Founding Father Matt Purdy's ghost, who blesses Fanny:

> They are the voices of thy children. They wait for you up the years and you must go [...] I steer you back into the mess of life because I was blinder than I knew.
>
> (1997, preface)

At the Old Vic, light faded finally as Fanny is rowed ashore to meet her lover waiting on the far bank. That this island off the Irish coast is also off the shore of Europe echoes the life of these individuals on the edge of public events, but nevertheless part of a wider history. Religious tolerance, rather than conflict is not an evasion but claimed by Barry to be an imaginative retrieval of a relative, who was not mentioned because she had married outside her culture, stating

> My father did not know much about her, because his father had never spoken about her, his own mother.
>
> (1997, Interview, Mick Moroney, p. 4)

Barry has elsewhere (my interview) acknowledged that on returning to Ireland in 1985 after some years in Europe, he felt none of the available identities of Irishness seemed to fit. 'Since I was now to be an Irishman, it seemed I would have to make myself up as I went along' (1991, p.v). Thus processes of memory, both private and public, are shown as important elements in the creation of cultural identity.

Although Ann Devlin's *After Easter* has a form apparently closer to realism than plays discussed so far in this chapter, it draws upon a family's memories, and includes accounts of visions as well as a revenant, while opening and closing monologues provide a heightened frame. Rooted in the political and ideological context of Northern Ireland, its central focus is Greta, whose breakdown is caused by post-natal depression and struggles with her complex cultural identity. Devlin, born and brought up in

Northern Ireland, had already won the Susan Smith Blackburn Prize and the George Devine Award for *Ourselves Alone* (1985, discussed in Chapter 5). *After Easter* was first performed by the Royal Shakespeare Company at their Other Place, Stratford, in 1994.

Set in 'the present', events follow Greta's collapse, from her incarceration in hospital to her sister Helen's flat, where her other sister Aoife has taken her, then back to Belfast. Here scenes take place in a nunnery, family home and the hospital where her father is fatally ill, returning to Westminster Bridge in London, and then an unspecified nursery or story-telling space. Greta's father, a Communist, and her strongly Catholic mother, who had struggled to keep their economy afloat, had not been happily married. Damage, including sibling rivalries, had thus been inflicted on their three daughters and their gay son Manus. Helena, a successful capitalist to spite her father, secretly sends money to an orphanage run by nuns, while Aoife seems the traditional mother of a large family, but is sexually and emotionally trapped. Greta's confusion of identity, intensified by her marriage to a Marxist Oxford academic, has caused her until now to refuse to return to Belfast.

Greta has firmly repressed all aspects of her Irish identity, especially her religion, which now returns to haunt her through a series of visions, as described to her sisters. She claims to have seen and heard the banshee, a death portent, although later implies it was another religious figure. Tensions within the family are interspersed with reverberations from the Troubles, not only in terms of the family's long-ago eviction from a non-Catholic area, but also through 'offstage' victims of violence in the Belfast hospital, damage done to Helen's car and the eruption of soldiers chasing Manus for going through a road block. When Greta steals a chalice from a church and offers holy wafers around town, Helen comments:

> In England they lock her up if she's mad, but let her go if she is political. In Ireland they lock her up if she's political and let her go if she is mad.
>
> (1994, p. 47)

Helen points out to Aoife that Greta is no madder than anyone else in the country. Ironically, this speech is soon followed by news that on her arrest Greta had written a statement about the need for immediate integration of schools – that is, an education system that mixed Catholics and Protestants. Meditating beside her father's coffin upon a confusion of identity, which she feels she shares with an illiterate man from Mayo and a Hindu immigrant child, Greta senses there are

> [...] no individuals, only scattered phrases and competing ideas which people utter to bewildering effect all the time [...]
>
> (1994. p. 59)

Only after her father as revenant leaves his coffin and speaks to Greta can she and her family begin to come to terms with the past, both personal and public, and the way it has affected their own present identities, while sounds of unrest are heard outside. Later, near Westminster Bridge in London, Helen and Greta throw the last of their father's ashes into the Thames, having thrown the rest into the River Bann. Sharing past feelings and the need to look forward enables Greta to go back to her family, as a baby is heard laughing. Finally Greta, as if reconciled both within herself and to motherhood of children from two cultures, tells a mythic story of origins to her baby, next to a chair left empty as for a traditional Irish story teller, the seannachie:

> [...] he took me to the place where the rivers come from, where you come from [...] and this is my own story.
>
> (1994, p. 75)

This use of a cultural device traditional in form and content is typical of resistant postcolonial strategies:

> a theatre praxis based on story-telling conventions foregrounds history not as a preordained and completed truth, but rather as a continually (re)constructed fiction which can only ever be partial [...], provisional and subject to change.
>
> (Gilbert & Tompkins, 1996, p. 137)

As indicated by Arrowsmith ((ed.) Brewster et al., 1999. pp. 140–2), Greta's re-reading is a means of 'self-location, self-identification and self-recognition', which transcends an essentialist view of national identity and is especially appropriate in a diasporic context. Myth and other traditional, ritualised theatre practices are also crucial within postcolonial performance (Gilbert & Tompkins 1996, pp. 53–105). Blurring of reality and fiction, transformations, repetitions, disruption of conventions of time and space, focus on the performative – or sometimes carnivalesque – body are especially linked with questions of community and efficacy. Although such performance is best suited to a theatre space which is not separated from the audience by a proscenium arch, it is claimed that, as a shared experience at its most effective, 'ritual drama aims to make the processes of participation' at a psychological rather than at a literal level, 'conscious and therefore potentially powerful as part of a larger communal project that images liberation from cultural oppression' (ibid., p. 65). Like the relationship between listeners and a traditional story-teller, drama incorporating ritual elements depends on shared emotional and cultural responses in terms of both authenticating and rhetorical conventions, rather than the appeal to reason and objectivity encoded in Brechtian alienating devices.

Finally this chapter returns to two plays that differently deploy some elements of ritual and carnivalesque strategies in their exploration of political history and myth: McGuinness' *Mary and Lizzie* (1989) and Vincent Wood's *At the Black Pig's Dyke* (1992). Both plays begin with different variations of myths of origin, and share some formal strategies and thematic concerns. Some, such as issues of border/liminality, the revenant, and sexuality, have been shown in this chapter to be linked to the splitting of identity. McGuinness' play follows the somewhat picaresque lives of Mary and Lizzie, two Irish girls who were, in fact, mistresses of Frederick Engels, friend of Karl Marx. First performed in the Royal Shakespeare Company's Pit studio at London's Barbican, it starts and finishes with a community of nameless women, including one pregnant one, living in trees at some unspecified, mythical time. These women speak in rhythmic, sometimes Gaelic and often opaque phrases, and use symbolic items such as a human bone, a bayonet and a stirring spoon. Although they are sisters, Mary and Lizzie – 'I'm utopian, she's scientific' (1989, p. 37) – are represented almost like a split self in the way they speak and act, always together. Loose narrative follows the orphaned sister's wanderings through both mythic and historical locations and events. Dispossessed, they encounter an old woman who says they will suit her son, a magical and anarchic priest who has been both Catholic and Protestant, and their Mother who has returned from the dead in baroque style to tell them to go to England, although it is no promised land. Her prophesies about disasters coming for Ireland are evoked by her companions' witchlike chants, and a dancing pig (a landlord) who sings:

> God protects the rich and the rich protect themselves
> The poor can go hang and the Irish go to Hell.
>
> (1989, p. 19)

After meeting a disenchanted Queen Victoria, they trace their Father to Manchester, and he joins them as they show Engels the industrialised city's horrors. Where Marx and Engels discuss the materialist view of history and capitalist economics, what seems rather heavy on the page may in performance be lightened somewhat by the sisters' often disruptive behaviour, and their double act of sexually pleasuring Engels (1989, p. 36). However, Jenny Marx quotes back to them derogatory extracts about the Irish from Engel's *The Condition of the Working Class.*

> Drink is the only thing that makes the Irishman's life worth living. His crudity which places him a little above the savage, his filth and poverty, all favour drunkenness.
>
> (1989, p. 40)

Countering this with a folk love song, the sisters, aware Marx and Engels as theorisers have a typical fear of the working class in actuality; say they will show them darkness which will arise from this fear. The last scene, partly in Russian, shows a grieving lost child, then all the women return to elemental chanting and magic ritual about the healing power of the earth to which sisters and mother now return – expressing mutual love, but aware that they have existed as but a marginal line on the borders of history. Contrasting discourses of masculine, public history and politics as opposed to a more private, feminised and mythic sphere almost in the Greenham Common mould does seem rather close to gender essentialism. The play's circularity, its fragmentation of time and emphasis on birth is also reminiscent of Julia Kristeva's concept 'Woman's Time', further discussed in Chapter Five. Nevertheless, the strong and evocative presence of these elements as embodied in ritual elements strengthens the more overt critique of exploitative economic elements in the colonial relationship between Britain and Ireland.

Whereas *Mary and Lizzie* has a clear historical basis, Wood's *At the Black Pig's Dyke* ((ed.) Farleigh, 1998), uses the device of a Mummer's play to convey border-crossing violence between North and South, Protestant and Catholic, thus linking present day assassins to those of the past. First directed by Maeliosa Stafford for Druid, and designed by Monica Fawley, it opens and closes with a myth of origin. Carried through as a quest narrative, it culminates with another vision of women as restorers of peace. The main female role, Catholic Lizzie Boles (nee Flynn) is played as an adult by one actress, but another plays her young self, her daughter Sarah and her grand-daughter Elizabeth as an adult. This splitting and doubling of identity further suggests and problematises the persistence of violence through time, across generations. Jack Boles, her Protestant husband, is played by the actor who doubles as First Hero, the Orange Knight, in the Mummer's play. Hugh Brolly, who has married Sarah, is played by the actor who also plays the Green Knight. Ironically, Lizzie's violent and jealous would-be lover is played by the actor who performs as the Doctor in the Mummer's play. The outer plot's complexities also draw upon the drowning of Hugh's ancestor by his jealous master, a nephew of Lord Leitrim, over a cross-class, cross-religion love affair, like that of Lizzie Boles. Further deaths – of Jack for being both shopkeeper and Protestant, of Sarah's son Sean and of her husband Hugh due to cross-border gunrunning, and Frank's shooting of Lizzie – are intertwined with repeated versions of the traditional Mummer's play. Further framing is provided by Tom Fool and Miss Funny, who act partly as narrators, and partly as commentators. In performance this Brechtian element is much less in evidence than the Mummers' stylised sinister movements as detailed in the comprehensive nebentext. Repeated often rhymed interchanges and folk songs contain ritual elements such as beating the ground with their sticks, dance, circular movements, and violent if ritualised murder. Traditional practices critique hate directed against identities feared as different. As Lizzie says:

> Men with masks, men with sticks, men with their mouths full of rhyme, men with their hearts full of hate, men with their minds stained with blood. Men to dance at a wake. Men to cry at a birth, men out searching their shadows.
>
> (1998, p. 33)

This chapter has explored a range of ways in which different theatrical forms and strategies have represented memory, history and politics. It seems that repressed elements of the colonial past are often embodied as revenants. Grand narratives and dominant ideologies, which tend to be associated with notions of unified subjectivity and realism, seem to be constantly interrogated by the split subjectivities and fragmenting techniques of many contemporary Irish dramas. Theatrical space, seeming somewhat akin to Kearney's concepts 'the Fifth Province' (Hederman & Kearney, 1977, p. 4), and ' "the radical imaginary' of Irish society' (1997, p. 69), may provide a liminal, heterotopic site, from within which hybrid postcolonial identities may emerge. The significance of women's role in reconciliation, which has arisen in some plays discussed, leads on to the next chapter's analysis of the representation of women in the Irish cultural context.

4 Madonna, Magdalen and Matriarch

This chapter explores contradictions implicit in the representation of women in contemporary Irish drama. First, it indicates the socio-economic and ideological context which informs attitudes to female identities, indicating points of similarity and difference in views held on both sides of the Border, with reference to the significance of the family and the growth of feminism. Second, after acknowledging formal and stylistic characteristics, which have been associated with women's performance and may have a specific resonance for postcolonial drama, some selective examples of the role of women as writers and directors in Irish theatre follow. Framed by reference to the representation of three female stereotypes in Murphy's *Bailegengaire* (1985), detailed comparative analysis then centres on two plays written by women and one by a man: Marina Carr's *Portia Coughlan* (1996), Marie Jones' *Women on the Verge of HRT* (1996), and Martin McDonagh's *The Beauty Queen of Leenane* (1997). These plays, all shown in London during 1997, were all directed by women, and all three deploy theatrical space as a site for the performance of desire. Discussion then examines women's perspectives on history, politics and family as staged in further plays by Ann Devlin, Christina Reid, and Elizabeth Burke Kennedy. Comparative reference to further plays by Carr, as well as Barry's *Our Lady of Sligo,* which problematise woman as mother and lover, precedes analysis of two male-written monologues, which critique religious and social factors that have damaged women.

Feminist analysis commonly places representations of women between polarised images – virginal Madonna and sexualised Magdalen. In the specific context of Irish cultural history, symbolic identification of women has been intensified both by the influence of Catholicism and by association with images of Nationalism. For example, Yeats' provocative play (1902) fuses such images of Ireland as the beautiful, young Cathleen Ni Houlihan with the suffering Poor Old Woman, the Sean Bhean Voch. Such traditions draw not only upon folk lore, but also on a romantic view of peasant life – like that associated with Anglo-Irish elements in the Celtic Literary Revival yet somewhat challenged by J.M. Synge. These, combined with a celebration of Irish rural landscape, especially of the West, thus stand

> [...] at the centre of a web of discourses of racial and cultural identity, femininity, sexuality and landscape which were being used in attempts to secure cultural identity and political freedom.
>
> (Nash, 1993, p. 44, qtd. Gray & Ryan, 1996, p. 180)

Iconography of woman as pure, yet paradoxically as Madonna-like self-sacrificing motherhood is also rooted in the 1937 Constitution of the Irish State through Articles that also recognised the family

> as the natural primary and fundamental unit group of society, and as a moral institution possessing inalienable and imprescriptible rights [...]
>
> (41.1.1),

giving State recognition to the woman who, by virtue of her life within the home, gave 'the State a support without which the common good cannot be achieved' (41.2.1). As Margaret Ward has indicated, even the role of militant women during the Irish State's emergence has been marginalised by male historians (1991).

Whilst this domesticising attitude – like De Valera's 1943 images of Ireland as a rural Eden (qtd. Beale, 1986, p. 20) – has been superceded to some extent since the late 1970s by more open discussion in the media of divorce, abortion, single motherhood and female sexual frustrations (*Irish Times*, qtd. ibid., p. 10), more recent feminist analysis suggests traditional views on gender roles persist especially in rural areas of the Republic. Despite the Irish state's increasing relationship with the European Economic Community and multinational companies, Mary Robinson's Presidency for one term of office from 1990, referenda on abortion and divorce in 1983/1992, significant legislation including the 1979 Health (Family Planning) Act with its more liberal 1985 Amendment, and the emergence of a strong feminist counter-culture,

> Irish women continue to live their lives in the shadow of traditional symbols of Irishness and Irish womanhood which are still enshrined in the Irish constitution.
>
> (Gray & Ryan, 1996, p. 186)

The 1980s were considered by feminists as a period of church and state retrenchment towards traditional values. Only in November 1995 was the constitutional ban on divorce finally removed by referendum ' [...] carried by the tightest margin in the history of the state (50.03 per cent)' (Ailbhe Smyth, July 1997). Women still travel abroad from the Republic if seeking abortions unless there is a clearly defined health risk.

The pluralities of Irish life – particularly the actual difficulties of Irish working class urban women on both sides of the border – are rarely reflected in the media, according to Gray & Ryan, with exceptions such as Roddy Doyle's gritty urban novels. Life for women in the North, where longer-established industrialised urban centres were affected by unemployment, especially after the shipyards declined, was economically and domestically difficult. Old practices through which employers (generally Protestant) rarely employed workers of a different faith, crop up in several plays discussed in the following pages, which also show the effects of sectarian divisions in housing, education and opportunity. Although political and religious differences between Northern Ireland and the Republic are clearly very significant – particularly in terms of some issues associated with sexuality and reproduction – problems associated with the representation of women have considerable similarities which are related to socio-economic conditions:

> As in the South, a religious fundamentalism and a conservative state contribute to an ideology which constructs women as subservient, where in the national order, women

> carry the main domestic and child-care responsibilities regardless of whether they are in the paid sector of the labour market or whether both partners are unemployed.
>
> (Beale, 1986, p. 67)

That the pressures of the Northern Irish question may have 'arrested the development of feminist consciousness particularly in relation to gender inequalities in the home' (ibid., p. 68) relates to Beale's view that partition did not give women in the North more scope. Further as Monica McWilliams suggests,

> The views of many unionist politicians on matters such as divorce and abortion, are virtually identical to those expressed by the Catholic Church.
>
> (in (ed.) Smyth, 1993, pp. 79–99)

Gray and Ryan also observe the media has blurred political and religious differences by emphasising images of women from either side of the conflict as passive victims of grief. Older mythic traditions and folkloric females from *The Tain* to the *Kiltartan* legends collected by Lady Gregory, are rooted in pre-partition culture, thus sharing heroines, heroes and events across recent borders. It is not therefore surprising that similarities in iconography can be traced in drama, often re-worked in tune with postcolonial literary concerns.

Encouraged by Mary Robinson's Presidency, 'centrally about imagining and constructing a new and much more open and complex image of Irishness' (Smyth, A., 1997, p. 45), and the later election of Mary McAleese to the Presidency, changes are taking place. In the South, empowering grass-roots organisations such as the Women's Council of Ireland has over 150 affiliated groups, while in the North

> the success of the Women's Coalition in the 1996 elections to the peace talks indicates the willingness of many women to cross sectarian divides and form real working alliances.
>
> (ibid., p. 49)

This willingness towards co-operation rather than confrontation is supported by Smyth's citation of twenty participating groups in the Belfast-based Women's Support Network. As in the South, wider employment prospects especially for middle-class professionals have been opened up over recent years by the Celtic tiger economy; so in the North in the late 1990s and early 21st century, the regeneration of Belfast as a consequence of the Peace Process is indicative of positive employment opportunities for women. Thus, often reinforced by the emergence of community support groups, women from different backgrounds are beginning to take an increasing role in public life.

From a postcolonial perspective, Asish Nandy (1993) argues that 'a history of colonisation is a history of feminisation'. Such passive behaviour, need for guidance, unruly barbarous and romantic qualities are 'all of those things for which the Irish and women have been traditionally blamed and scorned' (Meaney in (ed.) A. Smyth, 1993, p. 233). A challenge to such dominant ideologies, which trap and silence both women and the colonised, may be variously encoded in postcolonial performance. That gender discrimination is likely to be explored in 'combination with other factors such as race, class and/or cultural background', paradoxically suggests the

> metaphorical link between women and the land, a powerful trope in imperialist discourse [...] one which is re-inforced [...] by much post-colonial drama, particularly by male writers.
>
> (Gilbert & Tompkins 1996, p. 213)

Coincidentally, some characteristics claimed by a variety of critics to embody women's performance style and strategies ((ed.) Llewellyn-Jones, 1994), are similar to some of those claimed for postcolonial drama. These include: presentation of the self as split; deployment of the gendered body in space in ways that challenge the gaze of the dominant hierarchy (male, or coloniser); reclaiming of a place in history; reworking of canonical (male, or coloniser's) texts; fragmentation of linear narrative and in the case of the dominant (male or coloniser's) language, either subverting it or creating a 'new', often polyphonic, language; and may also intertextually include re-worked aspects of indigenous myths and cultural practices from a woman's perspective. Some of these postcolonial strategies may seem postmodern: the major difference lies both in their underlying ideological drive and the historical depth lacking in the surface/collage approach of postmodernism. A thorough discussion of the full range of feminist approaches from socialism to psychoanalysis cannot be given here, but aspects appropriate to different plays are indicated during this chapter. Plays written and/or directed by women are by no means necessarily predominantly feminist, and certain works created by men may nevertheless explore relevant issues.

Feminist analysis of Theatre as an institution in England and elsewhere has indicated its tendency to be structurally hierarchical and male dominated, hence the proliferation of women writers, directors and performers working in a flowering of predominantly fringe women's groups, widely considered to have happened in Britain in two waves following the 1970 Equal Pay Act and the growth of the Women's Movement. These groups dwindled during the economic restrictions of the Thatcherite years by the late 1980s early 1990s. Anna McMullan (in (eds) Griffiths & Llewellyn-Jones, 1993, pp. 110–23) has explored in detail the ways in which Irish women writers worked from 1958 and during this period. Citing a significant article 'Towards post-feminism?' in *Theatre Ireland* (No 18, April-July 1989), where a number of women writers, actors and directors were wary of being categorised as 'women's theatre' since this might 'ghettoise' them, McMullan considers lack of recognition of women's contribution to the field is indicative of 'the power of patriarchal values in Irish society'. These help to undervalue women's plays – less frequently published – or consider their often co-operative activities in fringe and community venues as less significant than those of established, more literary writers. McMullan mentions a range of earlier work, including plays by Teresa Deevey, and a Peacock Theatre production of novelist Jennifer Johnson's *The Nightingale and Not the Lark* (1980). Johnson's play was relatively traditional in form, contrasting with much of the more adventurous women's work in the 1980s within supportive fringe settings, such as Dublin's Project Arts Centre, or Theatre in Education teams, North and South. However,

Johnson's more recent monologues about traumas caused by the Northern Ireland Troubles – *Twinkletoes* (1993), *Mustn't Forget High Noon* (1989) and *Christine* (1989) (published together, 1995) – reclaim women's individual voices across the sectarian divide. Writing, as it were from outside both Irish communities, it is difficult to assess the full impact of later socio-economic developments already discussed in this chapter, but further evidence and my more recent interviews with some key women mentioned by McMullan and Philomena Muinzer (1987) can indicate their career trajectories into the 1990s and beyond, providing an admittedly selective insight into current conditions for women in Irish theatre.

Paradoxically, 'The idea of the female is one of the most prevalent themes in Irish drama, yet women are seriously under-represented in the theatre decision-making process' (V. White, *Theatre Ireland* No. 18. June 1989, p. 33). Garry Hynes, Lynne Parker, Marie Jones and (Mary) Elizabeth Burke Kennedy differently illustrate the growing power of women in Irish theatre. In 1975 Hynes, with Marie Mullen and Mick Lally, founded Druid Theatre Company, which initiated many plays discussed here. Despite winning a Fringe First in Edinburgh for her play, *Island Protected by a Bridge of Glass* in 1980, Hynes centred on directing. At Druid in the mid-1980s she championed Tom Murphy, with productions that transferred from Galway to London's Donmar Warehouse. Artistic Director of Dublin's National Theatre at the Abbey from 1990 to 1993, her visiting directorships include the English Royal Shakespeare Company. Returning to Druid in late 1996, she has since worked with London's Royal Court theatre, in respect of transfers and joint productions. Most notably, she discovered McDonagh through an unsolicited script, *The Beauty Queen of Leenane.* Her production went on to Broadway in New York, winning her a Tony Award, the first ever given to a woman for Best Director. She has also directed Marina Carr's *Portia Coughlan* (1996) for the Abbey. Her production *On Raftery's Hill* (2000) for the Druid is discussed in Chapter Six. Hynes, a dynamic director with a strong visual sense as well as an ear for the balance between tragedy and comedy typical of Irish drama, works with the Irish canon as well as new writers. Druid's policy is to tour Ireland, including even the islands, despite the problems of shipping materials. Hynes links her central interest in Irish plays to her birth in the 1950s, the watershed between the old and new Ireland. She says as she grew up,

> [...] all around me were the remnants and stories of an Ireland that was disappearing or gone. It was the need to explore that life that had gone before – and how we had got from that to this that interested me.
>
> (*The Guardian*, 03 July 2000)

While she has produced work by women such as Carr, and previously Geraldine Aron, Hynes does not stress feminist concerns per se. Exploring not only myths, but what is meant by Irish identity in the modern world, she is now a powerful figure in international terms.

Lynne Parker, born in Belfast and educated at Trinity College, Dublin, has worked with many companies, including Druid, Tinderbox, Charabanc and 7:84 Scotland. Interviewed as co-founder and Artistic Director of Dublin-based Rough Magic Theatre

Company in July 1997, she has since extended her reputation through visiting directorships at the Royal Shakespeare Company, tours, transfers and productions in England. These include work at London's Donmar Warehouse and Almeida Theatre, directing April de Angelis' *Playhouse Creatures* for Sir Peter Hall at the Old Vic, and *The Importance of Being Ernest* for Jude Kelly at the West Yorkshire Playhouse. Rough Magic received funding from 1984 onwards, from sources such as the Arts Council, the Government Department for Foreign Affairs and Dublin Corporation, as well as gifts from patrons. At the time of my interview Project Arts Centre was closed for rebuilding and temporarily housed at The Mint off O'Connell Street, but has now re-opened. Parker remarked upon the pressing need for more performance spaces to cope with proliferating independent companies in Dublin, struggling due to a lack of subsidy. Rough Magic's then policy was to produce three kinds of work: plays not done by other groups, reinterpretations of the classics, and full productions rather than just readings of new plays, to encourage new writers. From the first Parker claimed, 'Rough Magic resisted being pigeonholed, but preferred a broader approach, with ensemble work, drawing upon very diverse personalities, with the emphasis on good writing and good acting'. In 1997 the company consisted of three Dubliners, plus another three people who were one half-Scottish, one half-Northern Irish and one Swede. This pluralism and diversity is further evident in Parker's comment, 'The idea of a purely Irish Play is a purely bogus one'. Parker's directing style tends towards the visual. She spoke of heightened naturalism as a means of 'expanding into the world of the imagination'. Further, she suggested that, even when drawing upon the real, a play's text was like a musical score. Parker attributed to her uncle, the late Stewart Parker, a deeply musical sense of the actor in his writing. Among her successes was a posthumous production of his *Spokesong*, and she has won Dublin Theatre Festival Awards for her productions of his *Pentecost* and *Northern Star* in 1995 and 1996. In recent years she has directed plays by her colleague Declan Hughes, including *Digging for Fire* (1992) and in 1997 both his *Halloween Night* and Paula Meehan's *Mrs. Sweeney* (discussed in Chapter Seven). During 2000 she directed *The Comedy of Errors* for the Royal Shakespeare Company, but also *Down the Line*, a new play by Paul Mercier (whose work is discussed in Chapters Five and Seven) at the Peacock for the Irish National Theatre, where she is now an Associate Director. In 1992 Rough Magic initiated a writing awards scheme for women playwrights, funded by the Gulbenkian Foundation, the Art Council and the Northern Arts Council. This scheme is typical of Parker's career, which demonstrates, like Hynes, not only considerable creative talent worthy of international attention but sheer dogged determination. Her wide experience of directing classical and modern texts outside the Irish canon on the fringe has perhaps helped her to gain access to directing in mainstream 'dominant' theatres.

In contrast, Marie Jones, whose career as a director, writer and performer has been grounded in her work in community theatre, has toured widely and successfully across Ireland and parts of Britain, but has only relatively recently had access to London's West End, first with a short run of *Women on the Verge of HRT* (1997), discussed in this chapter. Later followed the outstandingly successful Olivier Award-winning *Stones in His Pockets* (1999/2000), discussed in Chapter Six, which transferred from London to Broadway.

Born in the shadow of Belfast docks, Jones was with five other actresses a founder member of the Northern Ireland Touring Company Charabanc in 1983. McMullan (1993, pp. 121–2) describes how these women, some from linen worker families, created their first piece *Lay Up Your Ends* about the linen workers' 1911 strike on the basis of local research. This collective method – not unlike the method used by the Joint Stock Company in England – also involved Martin Lynch as script co-ordinator, Pam Brighton as director, and a period of improvisation. Jones has sometimes worked with Charabanc and others on similar devising processes, or written scripts. *Somewhere Over the Balcony* (1988) was their last collaborative production, while problems linked with Jones' single-authored *The Hamster Wheel* (1990) – the only Charabanc play published – triggered her resignation as Artistic Director (Roche, 1994, p. 242). Publishers usually see group-written scripts as problematic, but some of Jones' own scripts have been published. Charabanc's philosophy was to reflect all aspects of the Northern Irish community, and with energy and satiric comedy generally drew faithfully on the lives of its people. Sadly it closed in 1995. Jones had moved on to found DubbelJoint with Pam Brighton and Mark Lambert in 1991. Based in West Belfast, its two-fold purpose, as expressed in theatre programmes, was to appeal to the Nationalist working class communities and to convey their preoccupations and vitality to the rest of Ireland. Its plays, usually written by Jones and directed by Brighton, tour to Dublin as well as Scotland and England. An early success that also went to New York was *A Night in November* (1997), winner of the TMA Best Touring Production Award (discussed in Chapter Seven). Jones has not abandoned collaborative and inventive work within a community context. Mic Moroney considered her Belfast Festival production *The Wedding* 'one of the most affecting pieces I have ever seen [...] although written by committee, acted by amateurs and directed by no fewer than four people, the script is nicely finished by Marie Jones and Martin Lynch' (*The Guardian* 16/11/99). This review, while hinting at elitist prejudice against community theatre, also indicates its strengths. In a bold extension of promenade performance audiences were bussed between two working class houses in Belfast, in which the Protestant bride and the Catholic groom, are respectively enmeshed in preparations as the audience move between rooms. Events in the Presbyterian Hall and a Lagan-side Hotel, with free bucks fizz and wedding cake, culminated in 'a hilarious striptease and trash-disco lift-off.' Lynch and Jones were also involved in Tinderbox's *Convictions* (2000) for the Belfast Festival in 2000 (see Chapter Seven). Hopefully, Jones' Broadway smash hit will not only gain her the wider reputation she deserves, but will facilitate both the further funding and artistic acceptability of community theatre.

(Mary) Elizabeth Burke-Kennedy, interviewed in my home (1997), suggested the huge resurgence of the Arts in the Republic implied 'as though for the first time the Arts have been put centre stage by Government Policy'. Not only EEC money but the influence of the outgoing Minister for Arts & Culture, Michael Higgins, himself a poet, had 'revolutionised the attitude to the Arts – and succeeded in getting funding.' Burke-Kennedy's career epitomises the move towards more image-based physical performance style, which has been developing more widely in Irish drama over recent years – as now evident in the work of Barabbas or Macnas. Co-founder of Focus with Deirdre O'Connell in 1967, Burke-Kennedy had largely worked with psychological realism, as in her first

Dublin Festival play, *Daughters* (1970), about two sisters and their dead mother, but she had also written children's plays. After seeing Peter Brook's *Company of the Birds* at the Avignon Festival, increasingly 'impatient with the restrictions of naturalism,' she began experimenting with 'ways of presenting theatre – over space and time, inviting the audience to play with you rather than give them a fixed angle' (Interview). Just as Brook remade African folktales, so she felt there was 'a treasure house of Irish mythology to be explored'. During the late Seventies and early Eighties, she edited and developed seannachie stories, through improvisation and music, using language, rhythm and movement to create images with bodies but without props or setting. Focus used its own material, but also Gogol's *The Nose*, Duras' *The Lovers*, Lorca's plays – using surreal puppets – and work from Kinsella's translation of *The Tain*, and a piece on shape-shifting for Cork Theatre Company. Burke-Kennedy left Focus in 1983, because the stage available was too small for the kind of physicalised image work that she now pursued, and her artistic policy had changed. In 1983 she formed Storytellers Theatre, which began reworking Victorian Anglo-Irish novels, such as Sheridan le Fanu's *Uncle Silas*, and was in 1997 looking forward to developing performance style further. A staging of Wilde's *Star Child* and other stories early in 2001 used animal masks, original music and puppetry to symbolise human characteristics. In 1988 *Women in Arms*, her reinterpretation of four major female figures in the Ulster cycle, discussed later in this chapter, was nominated for the Susan Smith Blackburn Award, performed by Cork Community Theatre Company at the Dublin Festival, and then at Tubingen, Germany by an Anglo-Irish Theatre Company directed by Ebhard (Paddy) Bort. Burke-Kennedy aimed to tell these stories through women's eyes, in a way that gave them contemporary relevance. Admiring Rough Magic and Paul Mercier's Passion Machine Company as examples of commitment to new audiences, she cited Barabbas and Blue Raincoat (of Sligo) as instances of physically inventive work. A then member of the Arts Council for over five years and Chair of Drama Sub-Committee, she pointed out the current significance of their support for the regions, especially Kilkenny, Waterford, Sligo and Limerick, and the importance of touring as some theatres outside Dublin were better equipped and designed for contemporary work. Some companies were beginning to take advantage of claiming a 'commissioning grant' for half the cost of publication, bearing the remaining cost themselves. Thus Burke-Kennedy's role seems typical of the way in which women in theatre of necessity combine creative and administrative skills in pursuit of the kind of profile and funding sources essential to the life of both companies and individual writers locally and internationally.

The career trajectories of these female directors are relevant to content, production context and history of the three contemporary Irish plays discussed as follows. All, set in a non-metropolitan context, that is, present-day rural Ireland – were performed both in Ireland and England during 1996/7. In exploring female identity – all three, *Women on the Verge of HRT*, *Portia Coughlan* and *The Beauty Queen of Leenane* are concerned with eroticism: female desire for sexual fulfilment under threat from either death or the fear of aging among other factors. Analysis includes the role of body, use or subversion of myth in creating a liminal space for the movement of desire, and the connection of eroticism with death through association with the grotesque matriarchal Hag as a figure of both warning and mourning. As previously suggested, theatrical performance provides a particular opportunity for embodying ideology and desire both through the actor's body in itself and as it moves through space. Further,

> [...] women's bodies often function in postcolonial theatre as the spaces on and through which larger territorial or cultural battles are being fought.
>
> (Gilbert & Tompkins, 1996, p. 215)

All three plays reveal the persistence of cultural battles around the stereotypical Madonna, Magdalen and Matriarch, which conflate the ideals of Catholicism and Nationalism.

Tom Murphy's *Bailegengaire*, set in 1984, was written in 1985 and directed for Druid Theatre by Hynes. Revived by London's Royal Court, directed in its studio space by James Macdonald (April 1997), it encapsulates the representation of these three key images of women. In a stylised west of Ireland country cottage, an aged, bed-ridden matriarchal Hag persists in re-telling what seems to be a seannachie-style narrative about the origin of the eponymous village's name, which 'means the town without laughter'. One grand-daughter, Mary, a prim virginal spinster nurse who has returned from exile to look after her mother, wants to force her finish the tale – which she never has. Dolly, the other grand-daughter, married to a battering absentee husband is a Magdalen figure pregnant by another man. Through intertwining narratives, it is gradually revealed that all three women have suffered emotional and sexual deprivation, while the repressed element in Mommo's story was the fatal burning of Tom, her other grandchild when both grandparents lingered too long at a pub. Painful and poetic, the play's blend of traditional and modern, birth and death, questions the power and persistence of these iconic representations. Mommo has also been seen as the epitome of

> [...] Mother Ireland, and her endless, and endlessly unfinished, story is an evocation of Ireland's buried children and buried history, of a historical grief that must be named and recognised before a country can be free of it.
>
> (F. O'Toole, in Murphy, 1997, p. xiv)

A final striking image as fading light foregrounds all three women together in the large bed central onstage – offers positive hope within an unconventional family structure:

> It was decided to give that – fambly ... of strangers another chance, and a brand-new baby to gladden their home.
>
> (Murphy, 1997, p. 170)

Women on the Verge of HRT, Portia Coughlan and *The Beauty Queen of Leenane* explore these three key images of female identity, that is, Madonna, Magdalen and Matriarch, in terms of the possible creation of a liminal space for female desire, which is connected with and subversive of aspects of Irish cultural mythology. Female desire is explored within three different spaces respectively: the mythic, the psychic and the literal.

Jones both wrote *Women On the Verge of HRT* and performed in it. Following success in Belfast, Dublin and touring, she obtained a London West End run at the Vaudeville – no mean feat for a piece with feminist implications. Although one critical response was 'The spirit of Shirley Valentine lives on!' and *The Guardian* (3 March 1997) ran an article entitled 'Is Your Mother in Love With This Man?', the piece has far more depth. Set in the present, through the phenomenal popularity of Daniel O'Donnell, a singer of light music and Irish ballads, it examines the position of two aging women. They are threatened with invisibility, the typical fate of older women today, since they are on the edge of menopause and hence facing a decision about whether to take hormone replacement therapy. Large numbers of middle-aged female fans flock annually to visit O'Donnell's Donegal home, queuing for individual contact and his mother's cups of tea. Opening large-screen documentary footage of such a visit, when both singer and fans co-operate with the camera, cuts to the onstage twin-bedded room in the Viking House Hotel opened by the singer in 1993, where two long-time friends are staying. Anna, played by Eileen Pollock, is dressed in sensible, unglamorous pyjamas, though she has bought a souvenir pillow emblazoned with the singer's face. Vera, played by Marie Jones, wears silky nightie and negligee. At first Anna seems to epitomise the domestic Madonna figure, while Vera is filled with rage that her financially successful husband has recently dumped her for a younger woman. Intending to live out her unfulfilled desires, she wants to be a Magdalen figure before it is too late. Initially the setting seems typical of a fourth-wall realist play, but as the women drink more and more – frequently sending for the tall, attractive but inherently physically clumsy young waiter Fergal, whom Vera eyes flirtatiously – the genre changes. Songs are interspersed throughout, some by the waiter (Dessie Gallagher when I saw the play) are somewhat in the O'Donnell mould. Those by the women reveal their desires and concerns – but both types are deconstructively placed. In a Brechtian fashion these songs reveal and further question through their metatheatricality the validity of the role of the popular singer and his songs as a fantasy outlet for women whose lives have been emotionally and sexually restricted by domesticity's demands. Soul-revealing discussion between the women is as painful as it is funny: the underlying failure of Anna's marriage is revealed, as is Vera's terror of becoming old, asexual and grotesque. Fergal eventually persuades both to join other women fans in a celebratory party, suggesting they meet him on the beach later as the sun rises. This suggestion is not a quasi-Shirley Valentine come-on, but turns out to be more magical. For the second act, the walls of the neat, claustrophobic hotel room have vanished and a magnificent open shore scene is suggested through back projection and lighting effects on the cyclorama. As the two women enjoy this beautiful spectacle, they hear the ominous howling of the banshee – the female portent of aging and death. Appearing as if by magic, Fergal as a shape-shifter role-plays various male and female characters in the women's lives: husbands, offspring, mistress, and so on. Through the process of interaction /role-play with these figures, both women are brought to a point of self-recognition – within this process rather akin to psychoanalysis/consciousness-raising. Transformed, as their final celebratory song/duet suggests, they will henceforth be stronger, more self-reliant and aware. Observation of a responsive, mostly middle-aged female matinee audience suggested that not only the humour but the depiction of aging struck a chord.

Apart from juxtaposition of two different female stereotypes, a significant absence – that which is most repressed – was the old grotesque Hag. Only her howling banshee cry is heard, though the shape-shifter waiter's mythic presence is perhaps linked with the banshee. The anguished voice of female desire is thus here, like that of the colonial subject performing a coloniser's text, both 'a locally embodied yet paradoxically disembodied voice' (Gilbert & Tompkins, 1996, p. 21). This banshee cry is the displaced voice of Anna and Vera, who are bodily present in the stage space. The question of their identity is problematised in the splitting that the women fear between what is perceived as their residually-desiring sexual present and its potential future annihilation. Patricia Lysaght's exhaustive study of this supernatural death-messenger emphasises the apparent absence of erotic traits, similar to the grotesque Hag of folk tales.

> Her general appearance is in the likeness of a very old woman, of small stature and decrepit form.
>
> (1986, p. 91, p. 95)

The banshee's characteristic wailing sound was considered to be a harbinger of death – thus here relevant to the notion that the play challenges, that menopause is the virtual death of female desire. Perhaps because the process of establishing subject identity has been linked with separation of the clean and proper body from the abject flux associated with its borders and orifices, menstruation – a sign of sexual maturity – was until recently, rarely mentioned in public. Although menopause has gained a higher profile through the debate on HRT, Anna and Vera embody a sense of abject self-disgust at this borderline condition. As Kristeva posits,

> what is abject [...] is radically excluded and draws me to the place where meaning collapses [...] On the edge on non-existence and hallucination, of a reality that, if I acknowledge it, annihilates me.
>
> (Kristeva, 1982, pp. 2–3)

It is only when they allow themselves to enter more fully the carnival spirit enjoyed by other female fans that their sense of bodily pleasures begins to re-develop.

Just as carnival is a period of relief from everyday life, so the Act Two shoreline at dawn provides a liminal, heterotopic site, appropriate for transformation. It symbolises not only the position of the women – poised between the ebb and flow of their menstrual cycles – but also a space of potential fantasy where desire is not confined within hotel bedroom walls, domestic restrictions or surrogate fan relationships. Fergal's Trickster-like, minimally signified shape-shifting, merely shown by ritualised exit and re-entry with additions such as headscarf, jacket or changed body posture, includes cross-gender roles. Embodying the power of imagination and desire to transcend the limitations of body-borders, it thus suggests individual potential to breach those restrictions placed upon her/him by social pressures and ideology. For the two aging women, it is a reminder that their energy and desires need not be limited by their bodies' boundaries. The magical re-presentation of past moments in the present breaks linear chronology, allowing the women opportunity for reinterpretation. Thus a traditional representation, the banshee/grotesque Hag, is changed from a figure of warning and mourning into a transforming and ultimately celebratory symbol. Jones' play creates a subversive, mythic space/time from which the women are re-born as neither Madonna or Magdalen figures, but as independent individuals, capable of seizing erotic moments in the face of time and death.

Portia Coughlan, written by Marina Carr as a commission for the Dublin Maternity Hospital and uniquely funded by Irish Women from home and abroad, was directed by Garry Hynes for the Peacock in March 1996, transferring to the Royal Court that May. Where *Women on the Verge of HRT* is ultimately an act of defiance against death, *Portia Coughlan*, despite comic elements, is a tragedy about a woman who seeks death as a way of obliterating her social identity. Set in a rural household on the eponymous heroine's thirtieth birthday, this play also disrupts linear time, starting Act Two with a tableau of Portia's dripping corpse, carrying on through her funeral and then returning in Act Three to events before her suicide, which chronologically would have occurred after the events of Act One. Again, Madonna/Magdalen stereotypes are challenged, while a grand-matriarchal Hag is unsympathetically presented. Portia has married Raphael, a rich, partially crippled factory owner, and has three children (unseen by the audience) whom she cannot bring herself to love. Torn by grief and guilt, possessed by the memory of her drowned twin brother Gabriel, whose ghost still appears to haunt her with his singing even after fifteen years, Portia whiles away her time by drinking with her one-eyed friend Stacia, and indulging in sexual dalliance with weak, unsuitable local men on the river bank. It gradually emerges that her relationship with her brother had been incestuous, and that they were also the unknowing product of another incestuous relationship between a half brother and half sister – 'same father, different mothers' (1996, p. 53). This latter secret was harboured by Blaize Scully, Portia's grandmother, the destructive matriarchal Hag, now confined to a wheelchair, who had been too proud to warn Portia's parents. At the funeral she evokes the inevitability of transgressive Portia's suicide:

> Where she war heading the day she was born, cause whin ya brade anumals ya chan on'y brin' forth poor haunted monsters.
>
> (1996, p. 37)

While Portia is represented as more Magdalen than Madonna in her rejection of her domestic role, her family includes other subversively drawn women. Maggie May, her aunt, visually a caricature of an aging prostitute, provides maternal support lacking in Marianne, Portia's own aggressive mother who, despite buying her a dress, rebukes her:

> Ya'd sweer ya war nevr taugh' how to hoover a room or dust a mantel. Bledy Disgrace, tha's whah y'are.
>
> (1996, p. 17)

> If ya passed yar day liche any normal woman ther'd be none a' this! Stop ud!
>
> (p. 19)

Carr considers Maggie May and her husband Senchil, as the humanistic point of focus amid monstrous experience (Interview). Grotesque elements are present, for example through Stacia, Portia's one-eyed friend 'The Cyclops of Coolinarney' with her range of different coloured eye-patches, and Grandma Blaize.

Where *Women on the Verge of HRT* drew upon authenticating conventions through its intertextual reference to popular and folk culture, *Portia Coughlan* draws upon rhetorical conventions through allusions to Shakespeare and the classics (see Kershaw, 1999, pp. 25–9). Nevertheless, Irish dreams of the American West are evoked through a tacky bar, the High Chapparrall, where Portia drinks. Upstage a rippling black sheet suggested the river, backing up three physical yet indicative spaces: the home dining table, the High Chapparall and the riverbank, scene of sexual assignations. In contrast, the psychic space of transgressive desire – 'The Fah dark river of the hart' (1996, p. 64) – which underlies and is in tension with the social restrictions of the rural community, is epitomised by this Belmont River, with its allusion to Shakespeare's Portia, another 'lady richly left'(*Merchant of Venice*). Portia's split-self provides a combined critique of both mother and Magdalen; through her fragmented identity she has a sense of abjection and loss, citing the love she made with her twin even in the womb:

> Buh ah thinche we war doin ud afor we war born, times ah clo my eyes an ah fale a rush a waher roun me, an above we har tha thumpin a me mother's heart an we were don know which of us be th'other.
>
> (1996, p. 63)

Gabriel, the revenant, appears in uncanny repetition, singing upstage, and his absence otherwise reveals the gap/space which desire cannot fill, that between Self/Other. Water here symbolises the flux of desire – a Lethe-like borderline, which Portia eventually crosses when she fulfils the suicide pact first made fifteen years ago.

For Act Two's opening coup de theatre, Portia's dripping slip-clad body, dangling high from a pulley, mourned by a Greek-like chorus of her friends and relatives, exudes:

> Dripping water, moss, algae, frog-spawn, water-lilies from the river. Gabriel Scully (her long-dead twin) stands aloof, on the other bank, in profile, singing.
>
> (1996, p. 31)

The heterogenous flux of her desire is one with that from her orifices and the river, and thus a further reminder of the state of abjection, which Kristeva associates with refuse and corpses as part of the process of separation through which identity is formed:

> These bodily fluids, this defilement, this shit are what life withstands, hardly and with difficulty, on that part of death. There am I at the border of my condition as a living being [...] such wastes drop so that I may live.
>
> (1982, p. 3)

Having been haunted by monstrous, uncanny desire, Portia has, paradoxically, transcended such limits. The play has explored her psychic space, through which, in Foucault's terms, she has sought

> [...] to return to the facticity beneath and beyond signs and symbols – death as the originary loss of the maternal body, of full unity.
>
> (1988, p. 105)

While the play clearly shows social pressures which restrict women – especially in rural communities – the suggestion that death may be the only escape, could be read as pessimistic, despite the use of radical and postmodern dramatic forms. In reply to my interview query about the play's potential feminist status, Carr replied, 'You can't write to an 'ism' [...] real feminists are quietly beavering away' (1997). She spoke about the savage world under civility in Portia's community.

> If you do something it comes out [...] Not to be wanting more would be far sicker [...] The journey of yourself is hard won.
>
> (ibid.)

Despite greater economic status, and thus superficial freedom, Carr suggested that underneath this, for women there is still the feeling of being, 'not quite up to scratch, a little fraction less than men'. Her plays are essentially language-driven, but with strong visual images on stage, which could be linked to Carr's subconscious awareness of choreographed movements in her reading of classical Greek plays. Although at that time she had lived in Dublin for more than twelve years, Carr's richly poetic work with its wide intertextual references, is permeated with echoes of the rural environment embedded in her consciousness since childhood.

Where *Portia Coughlan*, through contrasting social and psychic space, reveals pressures that tragically restrict female desire, *The Beauty Queen of Leenane* uses dramatic space and structure more conventionally, although it too focuses on body, eroticism and death. A joint Druid Theatre Galway/Royal Court (Upstairs) London production in 1996, after further touring in Ireland it returned to the Court's main house and was shown again in both Galway and the Royal Court (at the Duke of York's) in 1997 as part of *The Leenane Trilogy*. McDonagh, based in London although his family has now returned to live in Ireland, claims that he has not been influenced by other Irish writers, yet canonical aspects echo throughout his work, including the Beckett allusion in the title of part of the trilogy, *A Skull In Connemara*. Set in the present, the *Beauty Queen's* rural cottage, situated in a small Connemara town, was teeming with realist detail as specified in the nebentext, with a functioning Aga-type stove, sink, crucifix, and tea towel with tourist-style slogan 'May you be in Heaven an Hour Before the Devil Knows You're Dead'. Rain sometimes poured down outside the window. This fourth-wall design echoes the play's classic realism in its use of both time and space. Mag Folan, a grotesque matriarchal Hag in her seventies is intent on ruining her plain forty-year-old daughter/carer Maureen's last chance of romance with a neighbour's son, Pato Dooley, a labourer returned from England for a family party with visiting American relatives. The room's claustrophobia is broken only for a short spotlight scene backed by a small, almost abstract black and white urban 'English' backdrop (with red scaffold pole in the re-run), when Pato reads aloud a letter of proposal that he is sending to Maureen, suggesting emigration to America. Here the inscription of desire is traced only upon the body of Maureen – there is no mythic or psychic space onstage. Undertones of myth reside in the Hag herself, and folk music such as *The Spinning Wheel* sung by Delia Murphy, heard on the radio. Maureen's desire to be elsewhere, perhaps America, is spoken but not seen. Similarly, the disembodied television programmes watched by Mag and by Pato's brother Ray suggest other possible but not necessarily emotionally richer lives. The audience finds itself laughing at tensions between characters, which manifest themselves in non-sequitors, insults and Beckett-like boredom – not unlike a recent British TV programme, *One Foot in the Grave*, which portrays the elderly as irascible and manipulative.

Typical of the Hag's grotesque and abject qualities is her unpleasant habit of emptying her urine into the sink, pretending the stink is due to cats. Whining for her Complan, spitting it out when dissatisfied, rocking in her chair, watching TV and listening to the radio, driving her daughter to distraction through her continual orders, Mag hints at Maureen's previous nervous breakdown. Having managed to attend the Dooley's party, Maureen, transformed from Virgin to Magdalen entices the tipsy Pato to stay overnight. During their intimate talk he calls her 'The Beauty Queen of Leenane'. Next morning, flaunting her body in a silky petticoat and and embarrassing Pato, Maureen taunts Mag with details of their sexual activities – though later the audience learns that Pato had been impotent. Clues dropped throughout suggest Maureen has tortured Mag in the past. On later discovering her mother had burnt Pato's proposal letter which Ray Dooley should have delivered directly to her, Maureen burns her mother's hand on the stove, throwing cooking oil over her, which provoked audible audience gasps on the three occasions I have seen the play. In melodrama tradition similar cries of 'Oh no!' had been heard when Mag previously destroyed the letter. Whereas in *Women on the Verge of HRT* songs and popular culture were used to deconstructive effect, here comedy and melodrama – a significant aspect of the Irish dramatic tradition since Boucicault – are used to perpetuate stereotypical roles rather than critique the representation of women.

Nearing this skilfully structured drama's climax, Maureen recounts to her mother how she just managed to catch Pato before he caught the train en route to America, saying that she will soon join him in Boston. As she finishes, Mag tumbles from her chair to reveal a blood-stained and shattered skull – provoking further audience gasps, as Maureen intends to pass off as an accident her murderous deed with a poker. The last scene makes it clear Maureen's account of this reunion is just a fantasy as Ray Dooley informs her Pato is engaged to someone else. Ray too, though unaware, comes close to meeting Mag's fate. He points out as Maureen rocks in the chair:

> The exact fecking image of your mother you are, you sitting there pegging orders and forgetting me name! Goodbye!
>
> (McDonagh 1997, p. 60)

Ironically, as *The Spinning Wheel* is played too late as a birthday dedication for Mag, Maureen puts away in the offstage hall the empty, dusty suitcase she will never use to escape, and the empty rocking chair slowly stops moving as the lights fade to black. The implied inevitability of this cycle – Virgin to Whore to mythic, grotesque Hag – reinstates and confirms dominant patriarchal ideology. The brief play of female desire has been confined by the limited and literal domestic space of the set and reaffirmed by the inevitability of the linear plot and fatalistic repetition of circular spinning wheel song, a reiteration rather than a challenge to traditional folk values. Dreams of elsewhere have entered this space only through the economic lure of America and TV, although much humour depends upon the limited rural lifestyle that frustrates both Ray and Maureen in particular. The irony and hopelessness of Maureen's final message to Pato via Ray – 'The Beauty Queen of Leenane says goodbye' (ibid., p. 59) – evokes her identity as a mythic creature who never was, in contrast to her position as unbalanced if successful murderess whose sexual opportunities are now over.

The relationship of these three non-metropolitan plays to realism and to traditional tropes may be placed in terms of Kearney's three categories, which express different degrees of tension between the modern and revivalist tendencies in contemporary Irish writing. *Women on the Verge of HRT* is closest to the hybrid nature of 'mediational modernism', as

> [...] this [...] tendency cannot be strictly confined to either modernist or revivalist categories. It may be termed postmodern to the extent that it borrows freely from the idioms of both modernity and tradition, one moment endorsing a deconstruction of tradition, another reinventing and writing the stories of the past transmitted by cultural memory.
>
> (Kearney, 1988, p. 14)

Portia Coughlan, where intertextuality is less rooted in Irish folklore, could be considered as a formally – rather than ideologically – radical modernist piece as it resists the pull of tradition through its presentation of a fragmented subjectivity within a stultified community, reflecting 'the many and often competing elements that define postcolonial identity' (Gilbert & Tompkins, 1996, p. 231). *The Beauty Queen of Leenane* in its tendency to gravitate towards traditional stereotypes, dramatic style and structures, would be 'revivalist modernist'. The latter, written from an outside metropolitan perspective, is an example of Graham's 'new authenticity' (1999, pp. 22–5), encouraging a 'recreated heritage' – one that Fintan O'Toole considers conflates history with geography through a

> grid of stories designed to be plugged into a journey round Ireland for the tourist. Time and space become mixed up together. History is suspended in a commodified sense of place
>
> (1994, p. 40).

Thus, despite its financial and award-winning success, McDonagh's representation of female desire is restricted both by the theatrical space of conventionally-structured stereotypical comedy melodrama and by a commodified sense of Ireland as geographical space/place.

The site/sight of the body is in all cases linked with both eroticism and death through different associations with the abject and the grotesque Hag. Vivian Mercier's analysis of Irish comic tradition defines the grotesque in the context of Irish wake games, also relating them to another variant of the Hag figure, the Sheela-na-gig whose exaggerated genitalia he sees as indicative of the close relationship between eroticism and death (1991 edn., pp. 49–56). The Hag's presence in all three plays as a figure of warning and mourning is consonant with this notion, but it is most threatening and persistent in the *Beauty Queen of Leenane*, which is the most conservative both ideologically and formally. In the two plays written by women, the Hag is defeated through the subversive strategies of female desire rooted in either mythic or psychic space, which challenge the Madonna and Magdalen stereotypes, either through carnivalesque transformation or tragic transcendence.

Women's seizure of a more central space within politics, history and myth, from which they have been marginalised, may also start from the narrower sphere of domestic and personal relationships with which the three key stereotypes are associated. This movement is evident in selected plays by Ann Devlin, Christina Reid and Elizabeth Burke-Kennedy, which also support Maria M. Delgado's point that 'the feminization of Ireland into an idealised Mother Ireland [...] suits both coloniser and colonised' (in Reid, 1997, p. xvi). Kristeva's concept 'Womans' Time' is a useful strategy for reading how these plays challenge linear time through fragmentation, which differs from that discussed in Chapter Three, because it is linked not only to particularities of the post-colonial situation but to class, ethnicity, and social conditions primarily associated with female experience. Kristeva defines two kinds of time: '(the cyclical and the monumental) are traditionally linked with female subjectivity' (Kristeva (ed.) Moi, 1986, p. 192). First, citing earlier feminists' struggle to be inserted into history from which they have been excluded, she shows how this movement also worked for practical political ends such as equal pay, divorce, abortion rights and so on. Second, she defines the later struggle to voice the variety of experiences 'left mute in the culture of the past,' which inform a notion of identity that is not fixed, but 'exploded, plural, fluid' (ibid., p. 194). The way that such feminism 're-joins [...] the archaic (mythical) memory and [...] the cyclical or monumental temporality of marginal movements' can in these plays be seen in their attempts at 'insertion into history and the radical refusal of [...] subjective limitations' (ibid.). Devlin's *Ourselves Alone*, Reid's *Tea in a China Cup*, *The Belle of Belfast City*, *My Name Shall I Tell You My Name*, and Burke Kennedy's *Women in Arms* share some of these claims for female identity in their treatment of history and politics. Devlin ultimately concentrates upon a collective psychic space shared by women as opposed to male conflict, Reid reveals hidden significant connections between women's lives and these so-called male concerns, while Burke-Kennedy shifts to the foreground the under-represented role of women in cultural, mythic space.

As McMullan has shown, women's playwriting in the North has broadly something in common with the approach of those Loyalist and Catholic women who joined forces as 'The Peace People' for several years from 1976 under the leadership of Betty Williams and Mairead Corrigan (in (eds) Griffiths & Llewellyn-Jones, 1993, p. 119). *Ourselves Alone*, set mostly in West Belfast, echoes the slogan of Sinn Fein, with ironies that resound throughout. Devlin's first play, performed first at Liverpool Playhouse and then the Royal Court Upstairs in 1985, is concerned with women who live on the periphery of male activity. The play opens as Frieda is rehearsing a classic Republican song about Internment, while around her men are stacking mysterious boxes: 'I'm fed up with songs where women are doormats!' – 'I want to sing one of my own songs' (1986, p. 13). The women – sisters Frieda, Josie and their (common law) sister-in-law Donna – are in effect alone because their male relations and partners are embroiled in the complexities of the Northern Ireland situation, where violence was still being seen as justified in the attempt to claim (or oppose) the reunification of its six counties with the Republic. Their brother Liam's involvement with the Official IRA, then INLA, then the Provisional IRA while in Long Kesh puzzles the sisters as both politically opportunist yet confused (ibid., p. 22). An atmosphere of 'tribal' conflict makes separation of emotional and political loyalties virtually impossible, although the women want to resist it positively. Donna cannot quite relinquish her relationship with the often-absent Liam who is brutalised by his jail experiences, feeling she is now beyond the possibility of happiness. Josie, let down by a leading activist, becomes pregnant by an Englishman, Joe Conran. Although she had vetted him for the organisation, he turns out to be a double agent, whose betrayal is both political and personal. Josie's child becomes a sign of future possibilities. Frieda's Protestant lover John MacDermott is a teacher and committed socialist, but she realises that to be creative she must go to England: 'It's not Ireland I'm leaving it's him' (1986, p. 90). The play is not as negative as this outline may sound, nor is it entirely restricted by its predominantly realist approach. Not only does it give voice to women's perspectives, it finally shows their emotional, psychic solidarity with each other in a memory of a phosphorescent sea at night, when they had bathed naked together, laughing at their temporary escape from patriarchal tribal rules. The closing words 'How quietly the light comes' (1986, p. 90) are nevertheless tinged with optimism.

Reid's three plays share similar themes but more radical techniques. Roche (1994, p. 229ff) gives a detailed analysis of *Tea in a China Cup* (1983) with some reference to Kristeva's notion 'Women's Time'. Both this play and *The Belle of Belfast City* (1989) draw upon reciprocal female family relationships across generations, emphasised by circularity of structure, and female bonding rituals, such as tea-drinking. The earlier play traces three respectable Protestant women's struggles, through fluid flashbacks encompassing both World Wars and the present Troubles, which have affected their impoverished working class family. Sarah, the mother whose sickness opens and whose death closes the play, is in many ways complicit with the oppressive ideology of the male-dominated loyalist community, which is questioned by her daughter Beth – whose Catholic friend Theresa questions the ideas of her own community. The parallel lives and deaths of family males are epitomised in framed photographs of Grandfather (gassed in the First World War), his son Samuel (killed in the Second World War), and Sammy (for whom a return to Belfast might mean getting shot); all wearing British Army uniform. Family reactions to violent offstage world events are often literally framed onstage by the female gaze, culminating in Sarah's resistance to intimidation from all sides for staying in her home, now in a dangerous 'borderline' area. Socio-economic difficulties, religious pressures and injustices such as segregated schooling and employment are exposed through their effect upon the growing children. Female humour and resilience is celebrated, although on her mother's death Beth says, full of others' memories, 'I don't know who I am'. She takes away a cup from her mother's Beleek tea-service, despite reducing its resale value, because it embodies the female generations' shared concerns. Thus the often unspoken effect of linear, masculine history on women's lives is shown through the web of female relationships, which are stronger than officialdom.

The Belle of Belfast City, first produced at the Lyric, Belfast has more interspersed songs, which provide Brechtian commentary whilst manifesting the spirited, subversive energy of old Dolly, once a singer billed like the play's title. The past similarly intrudes upon the present, in events framed by Belle, Dolly's grand-daughter, watching 'as if hearing an often-told story re-created' (1997, p. 179), again through the web of female family relationships. Although Dolly's elder daughter, Vi, remains rather self-repressed, like Sarah in *Tea in a China Cup*, Rose, her much younger daughter, went off to London and success. The question of identity is complicated through Belle, Rose's daughter by a long-gone black American sexual partner, who calls herself 'an Anglo/Irish/Yank' (1997, p. 192). This is Belle's first visit, due to the family's reluctance to cope with racism in their community, and she is keen to learn about Ireland. Dolly comments, 'But as for Ireland, I've lived here all my life and I still can't make head or tail of it' (1986, p. 194). While Davy, a deaf-and-dumb youth with a mental age of ten, also foregrounds other questions of equal rights, the play explores issues related to women's freedom, and restrictions on both sides of the border. Rose says,

> We won't have many rights here if Jack and his gang get the independent Ulster they want. Their right-wing Protestant Church is in total agreement with the right-wing Catholic Church on issues like divorce and abortion, on a woman's right to be anything other than a mother or a daughter or a sister or a wife. Any woman outside those rules is the Great Whore of Babylon.
>
> (1997, p. 221)

A critique of extreme Protestantism's repressive ideologies is shown through the harsh sexual morality of Jack, Dolly's nephew, who was brutalised in his orphan childhood. Jack's sister Janet suffers from his threats because, having left her Catholic husband Peter, she had an affair with a young London man. Jack's connection with an extreme Right-wing British organisation, the National Front, is revealed through his covert sale of the small family shop to Bailey, an English businessman involved in the NF. The complexities of inter-community tensions in Belfast are made explicit through an anti-Anglo-Irish Agreement strike and rally at which Belle suffers racist taunts, and Dave, the innocent, is brutalised by the Royal Ulster Constabulary for acting in her defence. Ironically Davy is due to be supported in his claim for damages by Jack's contact, a lawyer who defends NF members. Belle is shocked that Jack allows Dolly's resuscitation from a stroke against her wishes. Vi finally takes Dolly away from Belfast at the open ending, as Belle sings the refrain 'May the Lord in His mercy be kind to Belfast' (1997,

p. 250). The use of 'Women's Time' and the potential multiplicity of female identities is here paradoxically strengthened because individual female roles are each played by a woman, but one man plays all male roles apart from Jack. Aware of their differences, women nevertheless persist in an emotional loyalty that is truer than that of the men, who are united only by a blind perspective: even Davy signs, 'No Pope here' (ibid.).

Similar tension between old male Protestant attitudes and younger female ideas about personal independence and Greenham Common-style solidarity is poignantly shown in *My Name, Shall I Tell You My Name,* originally a BBC Belfast Radio 4 play in 1987, later performed at the 1989 Dublin Theatre Festival. Old ex-soldier Andy, and his granddaughter Andrea onstage, apart, remember their past life together. She is drawing and he opens a box that contains her first drawing of him. As a child she used to recite a poem, which suggested that even when forgetting her name through shyness, she always remembered she was 'Grandad's pet'. The dialogue traces the development of their present utterly different positions, exacerbated by the fact Andrea has a daughter, Annie, with Hanif, a British half-Asian student from a middle-class family, who has been damaged by skinheads celebrating the British 'victory' in the Falklands. Andy remembers his participation in a parade commemorating the Somme, while Andrea, recalling her arrest at the Women's Peace Camp, Greenham Common, wishes that, like the brief Christmas truce between British and Germans in the First World War:

> There must be an hour, a place, where he and I can meet. A piece of common ground. A no man's land. If it's possible for strangers then it's possible for us.
>
> (1997, p. 276)

However, the last sound is a child's voice-over, singing the Unionist slogan, 'No surrender'.

The idea that women are essentially peaceful, passive creatures is challenged by Elizabeth Burke-Kennedy's direction of her play *Women at Arms,* performed by Storytellers Theatre at the Dublin Festival in 1988. In reworking the Ulster cycle of mythic stories, the *Táin Bó Cuailnge* it foregrounds four women – Nessa, Macha, Dierdre and Maeve – rather than traditional physical heroic prowess. Detailed nebentext focuses on actors embodying the texture of both landscape and elements, which are also captured by sound effects and what reviewer Derek West called an 'exquisitely painterly use of light'. Upstage was a large canopy – erected only twice to establish Conchobar's court – whilst on the raked stage five or six unevenly shaped boulders, suitable for actors to stand or sit upon, and props and costumes round the perimeter provided a minimal setting. As the (unpublished) script indicates, seven actors slip into a variety of roles, and form, with their bodies, a range of shapes from cloister to lake to burial mound, the great Brown Bull (of the myth's title), and even a tableau based on Picasso's Guernica. Such striking effects and stillnesses also gave prominence '[...] to the rhythms of the spoken word'.

Drawing thus upon a cultural myth, typical of postcolonial drama, this performance style also challenged the traditional trope of land and nationhood as possessed by the male. Here women, who do not fit key stereotypical roles, are celebrated for their beauty, craft, endurance. Maeve fights as a warrior queen to claim equality with her husband. Nessa, who becomes the mother of Conchabar, is a studious innocent, who eventually uses her sexual power to control the kingdom, encouraging cultural activities that made the Red Branch House famous. When Cruinnic forced his pregnant wife Macha to run in competition with horses, in her powerful distress while giving birth to twins, she curses all men who hear her scream so that in times of crisis all the men of Ulster down the generations will suffer. Dierdre's beauty and bravery in pursuing her love for Naoise shines through her defiance of Conchabar, in contrast to his deceit in allowing the murder of her lover. Her suicide, to avoid sexual possession by the king and the murderer, so affects Fergus that he leaves the Ulster court – thus playing a crucial role while in exile, in the story of Maeve, Queen of Connaught. Maeve's fury at being considered a 'kept woman' as her husband's possessions out-number hers by one powerful white-horned bull, causes her to attempt to own the great brown bull of Ulster. Her tactics drive her men into slaughter, while she throws doubt upon the prowess of Cuchulainn, Ulster hero, son of Conchabar. Nevertheless, after Maeve's defeat, her daughter stays with Cuchulainn. Throughout, all actors have shared telling these interlocking stories, but one – the story of the pig-keepers – is hinted at, yet never told. Finally, the story of escalating conflict between these two pig-keepers, one from Connaught and one from Munster, is told as a series of monstrous transformations, culminating in their becoming maggots, which are drunk by the two destructive rival bulls of the Cattle Raid myth. This epic ends with a tableau of conflict, which for critic Derek West had 'echoes for modern times'. Although content and performance style embody feminist strategies, the power of masculine myth is not entirely broken since the fate of these women is not entirely positive.

The problematic identity of passionate women whose strong feelings do not fit their socio-economic context, already explored by Carr in *Portia Coughlan,* is further evident both in her earlier work *The Mai* (1994) and *Beside the Bog of Cats* (1998). Both these plays and Barry's *Our Lady of Sligo* (1998) differently explore such alienation. Carr's plays draw upon both the power of Irish landscape and myths of origin. In *The Mai,* first seen at Dublin's Peaccock and in a later Tricycle London production, her daughter Millie remains on stage throughout, thus framing events partly as Brechtian deconstructive commentary but partly giving details of future and past. Down the generations, from the spinster whose one-night stand with a foreign sailor produced Grandmas Froechlain (still mourning her dead nine-fingered husband), whose daughter Ellen conceived The Mai out of wedlock, to The Mai herself and on to Millie and her illegitimate son, family history embodies strong, obsessive love that left women emotionally isolated. The Mai's determined drive to build a house at Owl Lake for her philandering musician husband Robert's return, has left little emotional space for her three children. Although a School Principal, she raised extra money by working all summer holidays in London. Millie bitterly remembers her mother gave love to a little girl there, instead of her own children. When Robert, with a new mistress taken after just over a year back at home, publicly shames Mai, it is implied that she drowns herself – in a situation parallel to the myth of the swans at Owl Lake. Robert is the only male character seen on stage, while The Mai's female relatives vary from grotesque, but highly subversive, Granny, to two interfering aunts and her restless unfulfilled sisters – one respectable and one promiscuous – try in their different ways to restrain her unrealistic dreams.

By the Bog of Cats, directed by Patrick Mason at the Abbey, has a poetic dimension not unlike the Greek tragedy, *Medea*. Here Hester Swane still waits for her long-lost tinker mother to return to her and the caravan where they had lived alone together near the bog. Like The Mai, she has an obsessive love for an unworthy man, by whom she has a daughter, Josie. Events centre around his marriage to a younger girl and attempts to get Hester to leave the house she had helped him to buy in the past, although they were never legally married. Thwarted in her schemes to win him back – there is a curious scene where his mother wears an almost bridal outfit to the wedding, and where the bride is further upstaged by Hester's appearance in her 'wedding' dress – Hester burns house and farming stock. Rather than relinquish her daughter, who has expressed a wish not be abandoned by her mother, Hester cuts her throat as a preliminary to her own suicide, because she does not want Josie to feel the loss she herself has endured. Earlier it is revealed that out of jealousy, Hester and her lover had killed her brother Joseph, born after her mother had abandoned her and returned to her father. They wanted the money left to Joseph by her father. These elements from the unrepressible past are revealed by this dead brother, another example of a powerful revenant. Other strange characters, such as the Ghost Fancier and the Catwoman – a grotesque Hag-like creature who drinks milk and eats mice – give a mythic quality to the play's poetic language and songs evocative of atmosphere and landscape. Hester's isolation is intensified by neighbours' insults about her marginal status as a tinker and her mother's implied whoring. As with *The Mai*, the circular effect of dysfunctional motherhood is stressed, but the imaginative intensity of all Carr's tragic heroines stands out against the drab pettiness of mundane, narrow-minded rural communities, thus embodying a critique of both female stereotypes and society.

Our Lady of Sligo, based on Barry's grandmother, shows the restrictions of a different community, which have hastened Mai O'Hara (nee Kirwan)'s death at 53 from alcoholism. Directed by Max Stafford-Clark for his Out of Joint company, it toured England – including the National Theatre's Cottesloe stage in London in 1998 – before going on to Dublin. The protagonist, powerfully played by Sinead Cusack, conveyed the sense that within Mai's decaying body some essence of her energetic younger self remained. Set in the hospital room where Mai lies dying, Barry's play evokes her past not only in Ireland but experiences of the colonies, as her family visits and her memories take shape. Upstage centre, a large Georgian-style window, indicated through lighting, conveys the passage of time, blurring past and present, imagined and real. Trapped in her bed, Mai's memories – not necessarily in chronological order – reveal how her independent spirit as a young university student was gradually eclipsed by a lifetime of disappointments and her tumultuous relationship with Jack. Roy Foster's introduction to the play text links her feelings with 'the disappointment of a whole class' (1998), that is, middle-class Catholics who did not find De Valera's vision of a frugal, rural new Ireland in the 1930's fitted them. Barry himself has regretted De Valera as 'a sentimental Stalin', who at that time closed down the possibility of Ireland becoming a 'European country in the old-fashioned sense' (*Guardian* (25/03/98). Jack had raised himself through British Army service, modelling himself on Ascendancy upper-class Protestants, but all in vain. 'Jack tried to turn himself into a sort of British gentleman, and by the time he achieved it, death and independence had erased his template' (1998, pp. 49–50). Having experienced the coloniser's life in Africa, Mai and Jack returned to Ireland, but their ill-matched marriage, based upon their love of dancing, drinking and dressing well, was doomed. Refusing at times the fact that she is married and has a daughter, Mai is haunted by past traumas, such as her sister Cissie's death in childhood – and that of her own baby son, for which Jack blamed her and her drinking. Mai O'Hara's daughter, Joanie, calls her 'Mammy of the gaps' (ibid., p. 11), implying their relationship was partial, due to her mother's own passionate preoccupations and lack of understanding. Joanie, now an actress at the Abbey, has memories of her parents' drunken parties, when she was dragged from bed to sing, and her mother's drunken vulnerability. She wonders whether her role was like the guards of the mythic Maeve:

> Did every broken mother need a sheltering child? And were you a dark queen like in the story? [...] And in the end I knew you were Our Lady of Sligo [...]
>
> (1998, p.44)

Thus Joanie had conflated her dysfunctional mother's suffering with that of the Madonna – and there is a sense in which Mai's life and death does embrace all three key female images. Typical of Barry's themes, the possibility of grace and redemption occur through traces of repetition, like the washing of a child in a basin which is repeated when Mai is washed in hospital. Mai's memories of favourite flowers appear just as Jack actually sends some, and she also has a vision that

> out of the savage nonsense of our life together that something might be gleaned.
>
> (ibid., p. 62)

The play begins and ends with Mai's memories of her long-dead Dada, who appears as a revenant. Finally, in a blue light upstage as Mai dies, we see and hear enacted a childhood moment when she, in a blue coat, is with her father. Nevertheless, the fragmented presentation of Mai's personal identity renders it as ambiguous and elusive as that of the nation and its withheld promises, shown in Jack's speech,

> Ireland, where is that country? Where are those lives that lay in store for us like ripe grain.
>
> (p. 23)

If motherhood is represented as problematic, so can old age reveal problems of patriarchal oppression, both through domestic and religious contexts. Whereas Murphy's *Bailegengaire* provided a positive closing perspective, two male-written monologues, McGuinness' *Baglady* (1985) directed by Patrick Mason at the Peacock and Dermot Bolger's *The Holy Ground* (1990) directed at the Gate, Dublin, by David Byrne, critique such negative effects. In the former, feminized only by the scarf hiding her hair, the elderly Baglady carrying a sack, walks around her space. Location is unspecified, but an audience could well be reminded of the similar fate that befell two sisters in *Dancing at Lugnasa.* The stream-of-consciousness flow of her speech is also littered with repetitions that become increasingly significant as the listener begins to piece together fragments of her story, which is broken in chronology as well as expression. The detailed nebentext suggests significant movements, such as rattling the big chain she carries and sometimes cradles like a baby. Her bodily gestures reveal her oppression. Later she deals out cards as if telling fortunes or a mythic story of a queen disguised as a beggar looking for her son. Gradually and painfully it emerges the man she fears is her respectable father, and that she was the victim of incest – compelled to drown her baby son after birth, possibly even when alive. Seemingly her mother may have been complicit in this secret. The imagery of her monologue is black, red, and white – like the memory of her house, blood, her red lemonade, her white child's underwear, the bread she eats, the devilish male memories and the black dog she fears. Constantly feeling dirty, she finally holds a white dress against herself, remembering how ' [...] a man in black came to her house and said she was a liar [...] and took her to the house of God [...] Women in black washed her' (1996, pp. 398–9). This reference to the church's role in confining unmarried girls and hushing up incest echoes the emergence of such scandals into the open during recent years. Following a repetition of marriage vows, the Baglady's last word is 'Drown' as she drops her mother's wedding ring, given to her by her father, into the river.

Whereas an unwanted baby is the centre of *Baglady,* barrenness lies at the heart of *The Holy Ground.* Set in a Dublin house after her husband Myles' funeral, Monica remembers the increasing narrowness of her married life, in a room dominated by a perpetual red light and an image of the Sacred Heart. Ironically the old TV is showing *Brief Encounter.* Originally 'Swifty' was quite a shy man, but loud in his love of watching and playing football. Courtship and early days in their house, which he had carefully prepared, had been happy enough. Once he sang 'The Holy Ground' with friends in a pub – with reference to Ireland, but potentially the football ground of his passion. When the marriage did not produce the children had Monica imagined, Myles was presumably told by the doctor this was his fault. Dismissal from playing with the football club intensified this humiliation. In increasing silence his thwarted masculinity became obsessed by participation in the (Catholic) Men's Fraternity, intensifying a narrowing morality and meanness. He forced Monica to help him write forged indignant letters about any potential modernising move in the church, such as divorce, birth control abortion and so on, and host society meetings at their home. Extremism led him to wander the streets with religious propaganda. Monica gradually reveals she thought she had succeeded in gradually poisoning Myles with rat poison – Warfarin used to thin blood clots. Dismissing her confession, doctors pointed out she had probably prolonged his life slightly by doses she had hidden in his food. In bitter irony, in her imagination she tells Myles he has not only taken away her youth

> [...] and left me barren, you've stole my gaiety and gave me shame, and when I die, I will die unmourned. But I could forgive you Swifty, anything except that [...] you had stolen my Christ away from me.
>
> (2000, p. 125)

As this chapter suggests, the representation of female identity is a complex issue, rooted in the socio-economic and cultural context of both sides of the border. Contemporary Irish plays draw upon a wide range of deconstructive approaches to the notion of women as Madonna, Magdalen or Matriarchal Hag, exposing the conditions that underly these stereotypes. Such factors are endemic to the link between gender issues and the post-colonial context, and some men as well as women do critique them. Here it would seem invidious to isolate certain performance strategies as 'purely feminist', although the dramas do deploy many styles and structures that have been claimed as such. Nevertheless, it is clear that certain talented women are effectively claiming a greater space for themselves in different roles within the Irish theatre industry. In contrast, Chapter Five considers the cultural implications of representations of male identity.

5 Myth and Masculinity

If an effect of colonisation is the feminization of the colonised, this then poses problems for their masculine identity, since traditional male qualities are associated with the dominant colonisers. Further, the impact of allegorical veneration of Ireland as Cathleen Ni Houlihan, and the idealisation of women as reinforced by religious morality, places males in a contradictory position. At one level, 'the need to ensure the hypermasculinity of the Gael' may have been the root cause of both Nationalist reluctance to encourage female suffrage, and its valorisation of family life (Cairns & Richards, 1991, pp. 130–1). Further, Patrick Pearse had no difficulty in

> eliding the martyred Christ and the martyred Cuchulain as a means of translating the defeats of historical time into the victory of mythic timelessnes.
>
> (Kearney, 1997, p. 118)

Riotous responses to representations of women by Synge and O'Casey previously mentioned also illustrate the Nationalists' need to perpetuate the myth of heroic masculinity, especially in memory of the Easter Rising's martyrs. O'Casey's *The Plough & The Stars* (1926) undercuts masculine posturing and heroics in that context. Attempts to develop an 'oppositional internal masculine representation of Irishness,' which stressed team sports, physical prowess and rugged rural men like those seen in Flaherty's film *Man of Aran* (1934), tended to ignore the reality of women's lives (Bronwen Walter, in (eds) Graham & Kirkland 1999, p. 80). Walter also suggests such stress on male physicality perpetuated the coloniser's attitude that the brutalised Irish male was 'incapable of self government', while feminine images of Ireland implied its need to be protected (ibid.). Deeply rooted racist attitudes in representations of Irish male migrants – for example, Roy Foster's selection from the earlier British magazine *Punch* (*Vol 3*, 18th March 1862), which describes them as something 'between the gorilla and the negro [...] the Irish Yahoo' (1993, p. 184) – proves the need to provide some positive male images was an urgent one. However, as Kearney's comments upon Gibbon's analysis of the allegorical use of female imagery suggests, such strategies

> [...] derive from a profound disjunction between expression and experience, outward sign and recondite meaning – a meaning which must be understood as part of consciousness itself under specific conditions of colonial persecution.
>
> (1997, pp. 230–1)

A somewhat similar disjunction seems to underly the representation of masculinity in much recent contemporary Irish drama.

Rather than emphasising the mythically heroic, men are often shown in reality as, in different ways, weak, even when a play centres upon masculine experience. This tendency echoes Kiberd, who, writing about the Irish literary renaissance, claims that

> The Irish father was often a defeated man, whose wife frequently won the bread and usurped his domestic power, whilst the priest usurped his spiritual authority.
>
> (1995, pp. 380–1)

His study of the relationships between fathers and sons tends to suggest these further assist in the emasculation of male power. After 1922 it was the successful revivalists, rather than the revolutionaries, who set the tone for domestically centred idealism in De Valera's mode (ibid., p. 293). Socio-economic studies and other commentators have emphasised problems that led to further migration either abroad or to urban centres, and thus affected the availability of women for marriage, especially in rural Ireland. Curtin and Varley's study of 'Marginal Men? Bachelor Farmers in a West of Ireland Community' ((eds) Curtin et al., 1987) gives a detailed account of economic and psychological factors responsible for the high percentage of unmarried landowners to that date. These include the system of land inheritance, increasing poverty of resources, extended family commitments and inertia, mixed with anxieties about responsibilities in relationships. Although rapid transition during the 1990s made the Republic, especially the cities, 'more internationalised, urbanised, cosmopolitanised and materialistic', at the same time

> Socio-economic poverty in regions outside Dublin became more pronounced, with the corollary of a growing illicit economy and crime [...] increasingly violent and drug related.
>
> (Pettit, 2000, p. 274)

Some plays analysed in this chapter explore such violence. Representations of masculinity in more urban Northern Ireland are further compounded by economic problems rooted in the past, such as unemployment arising from closing shipyards, exacerbated by linking of employment to sectarian positions. Some dramas reflect the violent frustration of men, for whom the male solidarity of sectarian organisations provided some sense of masculine identity. Although more positive aspects of economic change on both sides of the border – for example, references to international tourists, to new factories and new technology companies from Europe, America and Japan – may be reflected 'offstage' in some of these plays, the emphasis is on the performance of a problematised masculinity, especially in rural settings where younger women are now more likely to leave due to increasing independence and the possibility of jobs elsewhere. Even those plays that explore the urban scene's underbelly – the downside of the 'Celtic tiger' economy such as drug-taking, unemployment and unscrupulous building and so on – can be related to the construction of masculinity as well as location.

This chapter explores the relationship of myth to masculinity through comparative analysis of plays grouped under five themes. First is the representation of rural bachelorhood in plays by Barry, McDonagh, and McPherson. Second, after brief reference to Billy Roche's small town *Wexford Trilogy*, the urban context will be explored through work by Paul Mercier, Jimmy Murphy, Martin Lynch and Bolger. Third, the effect of violence bred from politics on representations of masculinity is introduced through brief reference to three plays from the early 1980s by Ron Hutchinson, Graham Reid and Lynch, which provides a context for more detailed analysis of Gary Mitchell's *In a Little World of Our Own* (1997) and *Trust* (1999). Fourth, comparison of the father's role in four recent plays – three by women – is followed by discussion of a fifth theme through two plays relevant to gay issues, Kilroy's *The Fall of Constance Wilde* (1997) and Gerard Stembridge's *The Gay Detective* (1996). Throughout, the relevance of a disjunction between the imagined and the real to concepts of masculinity will be examined with reference both to aspects of the postcolonial context and dramatic strategies.

Although Barry's *Boss Grady's Boys* (1986) and McDonagh's *The Lonesome West* (1997) are both about the by no means unusual demographic phenomenon of rural brothers living together, they are extremely different in structure, style and language, as well as attitudes to masculinity. Barry's Introduction to *Boss Grady's Boys*, directed by Caroline Fitzgerald for the Abbey, acknowledges he wrote it

> to repay a human debt to a pair of real brothers in a real corner of Cork, where I lived for a while in 1982.
>
> (1986, p. vi)

The nebentext indicates the forty-acre farm on the Cork/Kerry border should be implied by minimal properties, broad lighting effects suggestive of sky and mountains outside, with interior spaces shown by selective pools of light. Sixty-year-old Mick is both protective of and frustrated by his elder brother Josey, who is simple-minded and possibly suffering from Alzheimer's disease. Josey constantly repeats his anxieties – their horse out in the rain, a beloved dead dog, and his long-dead father's return from the fair. Mick sees Josey as a deep, dry well

> I throw stones into the poor man that echo a deep, lost sort of echo. I love him, I love his idiocy.
>
> (1986, p. 87)

Unrestricted by classic realist form, the play is atmospheric rather than driven by linear plot, and has a mediational modernist quality. Episodes evoke the brothers' daily life, shoeing the horse (conveyed through gesture), going to bed, or their different moments of reverie and dream about the past, in which their parents appear as revenants. Josey remembers fishing with his father; Mick recalls his dumb mother gardening, his own unfulfilled wish to emigrate, and their father's death. Some incidents – such as Mick's card-playing with neighbours, are embodied by actors, so the audience is uncertain about blurring between imagined and real events. For example the violent potential of the brothers' repressed sexuality is ambivalently shown in two different encounters with girls. Surreal touches include masks which lie in the brother's beds when they dream, and briefly revealed can-can dancers accompanying Josey's fiddle-playing. Allusions to Charlie Chaplin and Marx brothers' films, while also contributing to gentle comedy, imply the brothers' remoteness from modern, urban existence. Mick's memories of meeting Michael Collins suggest their forgotten rural lives are marginal to history, despite the idealisation of rural life intrinsic to the Nationalist agenda. A strong sense of powerful landscape and the harsh literal poverty of rural life permeates the play. Underlying tenderness, such as Mick's prayer for his dead father – 'Accept this most beautiful, most wayward father amongst other fathers' (1986, p. 115) – is typical of Barry's humanity, which is deeply emotional but never indulgent in its offsetting of pathos with the absurd. Despite winning the first Stewart Parker Award in 1988, the play's first UK premiere was ten years later, in Scotland during March 1998, when Lyn Gardner commented on its moments of happiness and poetic, 'luminous intensity' (*The Guardian*). Economics and unwillingness to break with the past underlie the brothers' acceptance of their mutual quasi-marital dependence suggested through the reality of small outbreaks of irritation, which are offset by their imagined thoughts.

Whereas *Boss Grady's Boys* celebrates the caring and loving potential of men, even within the confines of a narrow rural existence, the blackly comic *The Lonesome West* explores masculine violence endemic in such frustrating circumstances. Co-produced by Druid and London's Royal Court, directed by Garry Hynes in 1997, as part of the *Leenane Trilogy*, it won McDonagh the 1996 Evening Standard Award and George Devine Award both for Most Promising Playwright. Set in Leenane, Galway, in a typical farmhouse kitchen/living room, with table, basic kitchen sink and tattered armchairs, it shows bachelors Coleman and Valene as they return home from their father's funeral. Intense sibling rivalry is signified from the start by aggressively large 'Vs' with which Valene marks all his possessions, from whisky to rows of dusty figurines of the Virgin Mary which he compulsively collects. Apart from a short night conversation between Girleen, a pretty young poteen-seller, and local priest Father Welsh at a lakeside jetty (Scene Four) and the briefer moment (Scene Five) when the priest recites out front a letter he has sent to the brothers, action is confined to the kitchen. Although exaggerated in nature, events remain within the limits of realist form. Sibling rivalry about petty objects escalates from fisticuffs to death threats, to the despair of Father Welsh, whose parish contains other individuals driven to murderous violence through the restrictions of local community and family. These include Maureen, of *Beauty Queen* and Mick of *A Skull in Connemara*, other plays in the *Leenane Trilogy*. Details indicate the pettiness of local events, from the passion for vol-au-vents at funerals, a bored neighbour's suicide by drowning, to the prowess of the girls' under-twelves football team. Both brothers are sexually frustrated, taunting each other about their inexperience, especially through encounters with foul-speaking and sexually teasing Girleen – a potential Magdalen who harbours love for the priest. As in Barry's play, references to films emphasise the community's distance from the urban world, but here there is no mention of history. Gradually it emerges that Coleman had murdered his father by shooting him 'accidentally' in a squabble – an Oedipal struggle reminiscent of the failed patricide in Synge's *Playboy of the Western World*. When Coleman maliciously melts all Valene's plastic figurines in the new V-marked oven, Welsh deliberately immerses his own hands in the molten plastic as a sign of his own despair. In an attempt to turn his suicidal drowning to positive use, Welsh's final letter to the brothers via Girleen is a kind of emotional blackmail to stop them fighting. The last scene suggests at first that this ploy has worked, until it is revealed Coleman had knifed Valene's beloved dog in the past. Valene raises a knife against Coleman, and fighting escalates until Coleman shoots into fragments both the oven (a tricky explosive moment for the stage management) and all Valene's replacement china Virgin figurines. Despite worrying about the effect on Welsh's fate in the next world, the brothers carry on sparring once Valene realises Coleman has drunk away the House Contents Insurance payment. The last moments suggest that, although Valene is unable to burn the priest's letter, the brothers are likely to continue their cyclical violent bouts, interspersed with patches of repentance.

Although the satirical and comic drive of McDonagh's revivalist modernist play seems – as might his *A Skull in Connemara* – to exist on the basis of racist stereotypes like the 'ugly pugnacious ape-like cartoon figures of individual Irish men' discussed by Walter ((eds) Graham & Kirkland 1999, p. 78) and Foster (1993), experience of watching the play in London and then Galway showed that – especially in Ireland, the audience hugely appreciated the characterisations and local comic details, such as brand references. Further, the portrayal of an ineffectual yet well-meaning priest has obvious resonance with the Catholic priesthood's declining power. Nevertheless, the predominantly comic mode did not provide a consistent critique of the negative quality of masculine violence, or imply any positive solution to the deadly qualities of rural life as presented. The claustrophobic sibling relationship within a confined community echoes Beckett's similar couples in its intensification of absurdity and vengeful imagination. Even allowing for the possibility of reading the piece with postmodern irony, it still seems indicative of McDonagh's essentially outsider perspective, whereas Barry's concern as an insider is to present the dilemma sincerely, while implying that gentler, so-called feminine qualities provide a means of living with such difficulties. The 'weakness' of Barry's brothers has a moral strength that differs from that weakness which underlies the apparent physical strength of McDonagh's brothers. The question of authenticity in representations of Ireland – especially rural life – which these issues provoke, is discussed at length in Chapter Six. The Ireland that McDonagh represents, is, according to Conor McPherson, another young playwright who admires his skill, ' [...] not an Ireland I recognise, but it's great' (*The Guardian* 19 February 1998, interviewed by Fiachra Gibbons). Strongly appreciative audiences in Ireland suggested a cultural confidence that now interprets as celebratory some representations that would have offended in the past.

Paradoxically, while London-born and bred McDonagh's plays were welcomed in major theatres in Ireland, Dublin-based McPherson's works were initially rejected by such venues. Although earlier work was seen at University College Dublin and pub locations staged by Fly by Night Productions, and McPherson won a Stewart Parker Award for *The Good Thief* in 1994, it was the small Bush in London that brought him to the forefront through staging *This Lime Tree Bower* in 1995 and *St. Nicholas* in 1997. McPherson has also successfully written films – *I Went Down* (1998), an Irish road movie, and *Saltwater* (early 2001). Ian Rickson, Stephen Daldry's successor, staged *The Weir* in the Royal Court Upstairs (at the Ambassadors) in July 1997. Transferring to the Duke of York's, it became a long-running and profitable success, using a series of different casts, while the original cast went on to acclaim on Broadway as well as Dublin, Toronto and Brussels. Hence the Royal Court chose McPherson's *Dublin Carol* for their Sloane Square re-opening in February 2000, while his self-directed *Port Authority* presented by Dublin's Gate Theatre, opened in February 2001 at London's New Ambassadors.

The Weir, ostensibly a fourth-wall realist play set in a small bar in rural Ireland, in other words, north-west Leitrim or Sligo – illustrates the way contemporary Irish drama uses traditional tropes to extend beyond this form's potential limitations. Through drawing upon story-telling traditions somewhat in seannachie style, it explores the psychological implications of loneliness endemic in rural bachelorhood, touching in lightly its socio-economic context while exposing the genuine capability for human sympathy within the community – quite different from McDonagh's harsh treatment of rural masculinity. Originally on stage behind 'the iron', the setting intensified for the small Ambassador's audience – in close proximity to the actors – the feeling that they too were inside the tiny pub, smelling smoke from the fire. The compelling nature of both writing and performance, even on transfer to a more traditional proscenium staging at the Duke of York's, maintained a strong sense of intimacy. McPherson, interviewed by James Christopher (*The Observer* February, 1998), explained, 'I was technically obsessed with writing speech that wouldn't sound peculiar when it was spoken', and this skill is evident both in the apparently everyday pub dialogue and the gradual deepening implications of ghost stories told. Finbar, a successful, married businessman who had moved from the small remote village to the local town, brings Valerie, an apparently single lecturer who has moved into a local house, to the local pub, as part of introducing her to the district. The other three men are bachelors: Brendan, in his thirties, runs the pub and owns some land; Jim, in his forties, still lives with his infirm old mother, does odd jobs and has a bit of land; Jack in his fifties, runs a small garage. That a newer main road has now left his dwindling business on a side road is a metaphor for the way life has by-passed these lonely and unfulfilled bachelors, for whom the arrival of a new neighbour is an event. Their predicaments echo those cited by Curtin and Varley (1987), which presumably still persist.

As the evening progresses, with some friction between the bachelors and the twenty-pound-note-flashing Finbar, small details emerge about the local lifestyle, disrupted annually only by summer tourists whom the locals indiscriminately label as 'the Germans'. Under this superficially smooth surface run tense currents of emotion – just like the water of the Weir, photographed and hung on the pub wall long ago. The men's movement around the space shows tentative awareness of Valerie – Jim stays mostly rooted to his spot at the bar, Jack (especially when played by Jim Norton) moved in a lively way around the area, Brendan generally stayed safely behind his counter, and Finbar nearest to her. Ghost stories told by each of the older men contain unconscious echoes of their repressed emotions; significantly, the youngest does not yet need to tell stories. Jack's first story about the house Valerie has bought centres on mysterious knocking at its the doors and windows by fairies whose road has been blocked – suggestive of opportunities he did not take up in the past. Finbar's tale about the haunting of the Walsh family by a figure always watching from the stairs, suggests how he must have felt trapped by the community's close critical scrutiny, still persisting in the way he is judged even now. Jim's tale about being called with a friend to bury a man in a neighbouring village, who appeared to them as a revenant requesting burial near a dead child, does not imply that he himself was a similar pervert. It does suggest he is haunted both by the imminent death of his mother and his own buried sexual desires. These three ghost stories all draw upon oral traditions acknowledged by McPherson as linked to

> horror and violence in Irish culture. The famine for instance affected the way Bram Stoker wrote. It is also inspired by Catholic imagery.
>
> (*Observer*, 1998)

These qualities he feels are stronger in Ireland than 'a country which is stable and doesn't have a lot of mystery', and are thus typical of postcolonial discursive strategies.

Only when Valerie, who has separated from her husband, haltingly tells her own personal tale of hearing the voice of her drowned daughter on the phone are the men released from their inhibitions in awkward attempts to deny the 'truth' of their fictive stories as a means of consoling her. In a simple, touching moment, shy and inarticulate Jim (played by Kieran Ahern) moves from the safety of his position to take her hand, saying,

> I'm very sorry about what's happened to you. And I am sure your girl is quite safe and comfortable wherever she is, and I am going to say a little prayer for her, but I'm sure she doesn't need it [...]
>
> (1997, pp. 41–2)

After Jim and Finbar leave, Jack finds the courage to confess how his reluctance to move to Dublin cost him the girl he loved, and how at her wedding he felt

> [...] the future was all ahead of me. Years and years of it. I could feel it coming. All those things you've got to face on your own. By yourself.
>
> (1997, p. 46)

A small kindness done for him in his desolation was a sandwich made by a stranger: 'It fortified me like no meal I ever had in my life'. Finally Brendan finds his keys to drive Valerie home, implying she will be supported similarly by small kindnesses in this community of lonely people, which welcomes her. Thus re-working traditional story-telling provides a means for releasing repressed real emotions, and recuperates the genuine possibility of a humane community even within an economically deprived rural context. Its mythic function attempts to heal that kind of 'disjunction between expression and experience, outward sign and recondite meaning' (Kearney, 1997, pp. 380–1), which is evident in the men's ghost tales, but not in Valerie's story – nor Jack's honest confession. Yet the play also acknowledges problems of masculinity in post-colonial conditions intensified by the realities of daily rural life.

Secondly, plays about urban life, especially in larger population centres, suggest economics has a profound effect on notions of masculine identity. By way of transition between country and city, Billy Roche's award-winning social realist *Wexford Trilogy* (1989–1991), revived at the Tricycle, London (2000–2001), illustrates similar problems in small town life. Performed at Wexford Arts Centre in 1986 under the title *The Boker Poker Club*, it was re-titled *A Handful of Stars* and seen at London's Bush in 1989, then produced for TV by Initial films. BBC 2 showed all three plays in 1993. The first two plays are set in essentially masculine environments – a seedy pool room and a betting shop – although Elaine, a key figure in *Poor Beast in the Rain* (1989), works in the latter with her father. Both plays highlight small town-claustrophobia. Roche himself describes Jimmy Brady in the Afterword to *A Handful of Stars* (1992 edn.) as

> a small-town rebel who refuses to have his wings clipped or his tongue tied, who refuses to swallow the bitter pill of convention [...] he goes hurtling towards his own self-destruction.
>
> (1992, p. 35)

Jimmy, influenced partly by films he has seen, considers himself a notable local tearaway, swaggering in the non-elite part of the pool hall about his petty crimes and success with girls. Describing of his flight across roofs pursued by local policemen, who helped him as he nearly fell, he claims to be 'a nuisance and a danger to the public' and ends, 'The town has it off by heart. I'm famous sure'.

A dislocation between his self-created myth of heroic masculinity, and the reality of his pub fighting, relatively poor pool skills and petty thieving culminates in his eventual capitulation to arrest after a foolish episode with a gun. Nevertheless, Jimmy's cheeky, playboy confidence is more attractive than the beaten-down quality of the other males, who show rough loyalty and grudging admiration of Jimmy's wildness until he goes berserk when jilted. It is never made clear whether his one happy memory of parents dancing together was real or imagined, as illusory as the film rebels he aspires to be like.

Poor Beast in the Rain is described by Roche as 'a rainy day sort of play which is held together by an ancient Irish Myth as Danger Doyle returns like Oisin to the place of his birth' simply to see his old mates (1992, p. 188). Doyle, with a youthful reputation not unlike Brady's had run off with Stephen the bookie's wife, Eileen's mother, ten years ago, on the day of the All Ireland Hurling Final. During Doyle's long absence, his youthful tearaway days have been remembered and embroidered, especially by his old pal, Joe. Discrepancies between these memories emphasise the disjunction between the experience of these small-town misdemeanours and the mythic image of Danger Doyle the rebel. Doyle tells Molly, a cleaner who had once loved him, that like the mythic hero Oisin, who had been warned by his wife Niamh not to get down from his horse on returning from Tir na nOg to Ireland, he too will be 'turned into an auld fella [...]' if he climbs down from his imagined status (ibid., p. 117). The sense of being left behind by Doyle's imagined success is emphasised because local men are all in different ways unfulfilled. Yet getting away from Wexford is not entirely positive, as Doyle tells Eileen her once-vivacious mother is now seriously depressed. Local men are disgruntled by Doyle's claim to free them by revealing the mundane truth of his life (ibid., pp. 119–21). Hoping now to be forgotten, Doyle leaves – 'I swear to God I don't have what you seem to think I took from ye' (ibid.). Although disjunction between dream and reality has been exposed, no positive solution for small-town masculinity is suggested.

Belfry (1991) shows Artie O'Leary's progress from shy, repressed sacristan, to sociable card-and- snooker-playing confident man, as a result of his affair with Angela, a married woman who seems keen to clean and bring flowers to the church. Unlike Roche's earlier plays, the linear progression of events is disrupted. Artie addresses the audience in significant explanatory moments, thus breaking the realist frame. Other ineffectual men include Father Pat, temporarily 'off the gargle', and Donal, Angela's husband obsessed by his handball skills. Artie's ill and over-dominant mother, something of matriarchal Hag, is offstage. Artie embodies so-called feminine and masculine qualities – his concern for Dominic, a disadvantaged mischievous altar boy, is temporarily obliterated when he beats him, mistakenly thinking he has revealed the affair. Despite various unfortunate events, emboldened by 'a hidden reservoir [...] tapped' (ibid., p. 180) by habitually unfaithful Angela, Artie develops a more mature, pragmatic identity. Disjunction between the expression of romance and the actuality of small-town living seems here to have provoked positive results, Artie's oddyssey thus has a mythic function.

As in *The Wexford Trilogy*, social realism seems to be the dominant form used by many writers of the urban scene. Shaun Richards (in (eds) Graham & Kirkland 1999, pp. 102–3) supports both Kiberd and O'Toole's view, that in rejecting rural identities typical of the past, an emergent new writing about the 'heroism of urban life' was crucial to emerging new Irish cultural identities. Paul Mercier, writer, director and founder of Passion Machine (1984), is among those concerned with the realities of life in large urban centres and their impact upon young people – represented, for instance, in his notable trilogy about Dublin; *Buddleia* (1995), *Kitchensink* (1997) and *Native City* (1998). His work, which draws on different performance styles in celebrating the energy of working-class life, especially its youth, is discussed in Chapter Seven. His *Home* (1988) does show solidarity between males – in support and survival techniques given to Michael, an innocent young man up from the country to learn hotel management, by Valentine, a labourer living in the same Dublin apartment block. Martin Lynch and Jimmy Murphy also cover the underside of urban life in Northern Ireland and the Republic respectively. Lynch, a committed socialist playwright, has worked extensively in community theatre in Belfast since the 1970s – including co-operation with Marie Jones. *The Stone Chair* (1989) written about and for the people of Belfast's Short Strand area was a particular success ((ed.) Smyth, 1996, p. 10).

Lynch's earlier *Dockers* (1981), performed at the Lyric, is set in the Sailorstown district of Belfast during 1962. It provides a socialist perspective on the psychological and economic effects of casualisation of dock labour on working-class male identity at a time when there was one union for Protestants and one for Catholics – headed from Britain and the Republic respectively. Despite a tendency towards over-didactism in some dialogue, the play depends on contrasting main characters, John Graham and Buckets McGuinness. The former is a principled family man, elected onto his union committee;

the latter a cheerful, lazy ne'er-do-well, forever borrowing money and spinning tales, evading his wife and children. At that time, the union books rarely opened to new members, thus restricting many men to the mere chance of being taken on as boats docked. Two Catholic brothers – one with a pregnant Protestant girlfriend – are seen fighting each other for this opportunity. Failing to rally much support for the International Workers March on May Day, John's complaint echoes the disjunction between experience and expression:

> They'd rather march in their hundreds on St. Patrick's Day commemoration a friggin myth, than march as part of a working class movement through the streets of their own city. It's the same story in Ireland. Socialism versus the Saints.
>
> (1996, p. 56)

John's idealism in looking beyond sectarian divisions to international unity gets him temporarily expelled. Insisting on singing 'The Red Flag' (written by Irishman Jim Connell), when asked to join in pub celebrations, he declares,

> Even your bigoted minds give way to the singing of a Protestant song. It's all very well to play at being trade union leaders, but when the true concept of Labour is raised our leaders are terrified.
>
> (ibid., p. 82)

Others ignore his beating by so-called union brothers, lest they be ostracised by the foreman who picks the workers each day. Just prior to the Troubles, it seems for men the struggle to find work in a limited economy was a more immediate concern than sectarian difference. Paradoxically, feckless Buckets, somewhat akin to the Trickster role in postcolonial literature, seems more appealing than the idealist father, so that the comic almost outweighs the play's tragic elements. Masculine violence is never far away, and even John Graham shows timidity about the prospect of encouraging his intelligent son to find social mobility through education (ibid., pp. 64–5).

Jimmy Murphy's more recent plays, *Brothers of the Brush* (1993) and *A Picture of Paradise* (1995), give a very bleak view of the present boom in Dublin. The former, a Peacock production winning Dublin Festival Best Play and Stewart Parker Awards (1994), was later produced at the Arts Theatre, London. It shows a reversal of those socialist values expounded by painters in Robert Tressell's novel, *The Ragged Trousered Philanthropists*, dramatised by Joint Stock in 1978. Here, in a realist context, three male painters are employed by Martin, a Dublin entrepreneur seizing full advantage of new opportunities to gut and superficially renovate old city buildings by taking on at low rates men whom he knows are drawing the dole. With high unemployment, he can easily replace them with others or trainees on government programmes – he is not interested in getting skilled tradesmen. The play contrasts Lar, a reliable family-centred worker with his old friend Heno, recently returned from nine months in London. Despite his careless work, Heno threatens to inform the union of their working circumstances. When Lar is made 'foreman' pending a longer-term job, Heno persuades

Jack, Martin's elderly uncle who does all the unpleasant tasks, to 'go on strike'. Jack laments,

> Was a hard man, me! Anyone talk to me like that when I was young and that's what they got! [Makes a fist]
>
> (1995, p. 40)

Both Jack and Heno think Lar's concern for his family, which prevents him from wasting time and money drinking, is a sign of weakness (ibid., p. 48). Heno claims, 'He made a man out of me, my father' (ibid., p. 66), boasting his father earned respect through drunken sprees. With a cynical grip on 'any reality still out there' (p. 21), Heno uses blackmail to become foreman, so Lar is marginalised into short-term work. Murphy's characters are not written so the audience can identify with one position more than another; each has both weaknesses and dreams. Under economic stress both brotherly solidarity and loyalty to the firm seem worthless – and masculine self-interest seemingly triumphs over family values.

A Picture of Paradise, also seen at the Peacock, provides an even more downbeat impression of urban life and its impact on male identity and the family. Murphy aims to show 'The lives behind one of the many faces now populating the alleys and doorways of our cities', and the price to be paid for the rapid expansion of Dublin ((ed.) McGuinness 1996, pp. 170–1). A grey/brown set of apparently concrete walls gave a grim impression of both outside and inside the housing estate, its stair-wells redolent of piss and discarded drug paraphernalia. Here Angela, her son Declan and husband Sean attempt to squat after losing their home due to arrears, joined by The Lord, an ex-drummer who raises money with Sean by 'parking cars' in an obscure place near the city centre. Badly let down by Sean's promises over the years, Angela is by no means an idealised mother figure. Her gambling obsession has been largely responsible for their eviction as a result of a scam, and later it transpires she turned down alternative housing away from her access to bookies, moneylenders and so on. Dragged from school before his Leaving Certificate examinations, Declan has saved money to try to get on a computer training course, despite his tatty clothing, and struggles to keep the peace between his parents. Bitter comedy resides in the way their pathetic possessions are carried up to the flat in Act One – and cast out, wrecked, by (unseen) others, who seize the squatted flat for their own people as Act Two starts. Drinking thwarts Sean's attempts to contact those who might help him regain work as a chef. Even worse, The Lord's drinking after a band gig had led to the death by fire of his young family. Unlike several plays in which fathers jealously succeed in damaging their sons, when Declan fails his interview he refuses to give up his savings, leaving for Brussels to get bar work with a friend. The Lord and Angela drink amid the debris while Sean staggers off to their old home's shed. The 'Picture of Paradise' – a kitsch tropical seaside scene, which had hung in their home for 20 years is shattered like the marriage. This picture embodies the disjunction between the dream and the actuality of family life on the urban breadline, which takes away effective power from men – but also destroys women.

Bolger's perspective on working class life in Dublin is even bleaker than most, while his drama is more adventurous in form and technique. Already known as novelist and poet, Bolger's play *The Lament for Arthur Cleary* (1989), first performed by Wet Paint at Project Arts Centre for the Dublin Festival, won Samuel Beckett, Edinburgh Fringe First and BBC Stewart Parker Prizes. Its themes include notions of masculinity and heroism, urban deprivation, migration and return – all linked to questions of cultural identity and the wider European context. Murray suggests it is a re-working of an eighteenth century poem, beloved by nationalists, about a Catholic killed under Penal Laws on returning to Dublin after a period as a Hungarian Hussar (1997, p. 242). Both main roles – Cleary, aged thirty-five and his eighteen-year-old girl friend Kathy – are played by individual actors, while two males and a female play all other roles, including the key roles Frontier Guard, Porter and Friend respectively. The nebentext suggests simple staging, with a small box stage right, a barrel and a long thin platform far right, while sound and lighting effect throughout, combined with banging of sticks and occasional use of masks by the three doubling actors, enhance the surreal atmosphere. Kathy's voice is sometimes recorded, around her live words, especially when she sings, adding to the sense of a dislocated individuality. Her lament for the already-dead Arthur opens the play, which is typically disruptive of linear time, setting up a disjunction between dreams and harsh reality. Set in a kind of frontier post, not unlike those through which Arthur has travelled when working in Europe, it emerges that this spot, like the classical river Lethe, is the borderline between life and death. Within this heterotopic space, Arthur recalls past events and the course of his relationship with Kathy is played out. Four times during the play an interchange between the Frontier Guard and Arthur is replayed, including a joke about Irish identity: 'Ah. Irish. Irish. Boom-boom! Eh!' (2000, p. 5). When Arthur asks which side of the border he is on, the Guard replies,

> What difference to you Irish. I see you people every day, you're going this way, you're going that way, but never home. Either way you're a long way from there.
>
> (ibid.)

The exchange takes on an increasingly sinister quality, until Arthur realises he has to 'let go' of the country of the living to which he can never return. Gradually the love story between Kathy and Arthur is retrospectively revealed, against a background of economic difficulties and squalid surroundings on estates where unemployment and drug addiction are commonplace, and drink and clubbing help to obliterate disappointment for those not caught up in the economic boom. This point is underlined by a satirical speech given to a politician who smoothly addresses fellow European ministers about the

> [...] young people who are to Ireland what champagne is to France! Our finest crop [...] For export to your factories and offices.
>
> (2000, p. 6)

Drawn back by Dublin after working in Europe for 15 years, Arthur, whose spirited Mother though poor had been noted locally for her hospitality to anyone needing support, finds much has changed, including the demolition of his old haunts and the local iron church. His mother's old flat is now occupied by the aggressive Deignan family. He cannot find himself, nor his mother's spirit, in the new, even harsher environment – except in the company of Kathy. She realises that, unlike her father, whose kind

> [...] were never taught how to show grief' through masculine pride, even in unemployment, Arthur will teach her 'how to breathe.
>
> (2000 p. 23)

His wider experience of working with those from other cultures, including Turks, the sense of freedom epitomised both in his motorbike riding and an ability to find pleasure in the city's hidden places despite living on the dole, appeal to Kathy, although she would like to go abroad. Arthur's attitudes contrast with the mundane:

> What's real life, a clean job, pretending you own some mortgaged house on an estate, death from cancer at forty.
>
> (2000, p. 41)

The lovers are watched constantly by their neighbours, because they are different. Arthur's principles lead him to refuse to co-operate with Deignan, the local drug baron, who claims what money there is around the run-down estate is due to his organisations. Arthur's murder in an alleyway while Kathy sleeps is shown symbolically through lighting and sound effects, as well as strong though stylised physical movement. Finally he learns from the Frontier Guards that although Kathy mourned him 'begging you to haunt her' (2000, p. 67), she eventually had other children – 'She taught them your name like a secret tongue' (ibid.). While examining the problematic urban context, this play suggests the possibility of a masculinity that is not afraid to show tenderness, is aware and sympathetic towards cultural difference beyond Ireland, and has an honourable bravery in challenging corruption and violence.

The third theme, the relationship of violence to masculine identity in an urban context associated with political conflict, is inevitably most evident in plays about Northern Ireland. Especially when confined to a prison cell, the stage presence of the derogated (humiliated and damaged) body 'becomes a sign of the political fortunes of the collective culture', conveying the disempowered body of the colonial subject (Gilbert & Tompkins, 1996, pp. 221–2). Following the foregrounding of Republicanism through the period of the Hunger Strikes, an interest in examining the Loyalist position emerged in drama. Two typical examples from the earlier 1980s, a period which Pettitt (2000, pp. 235–9) considers to be associated with relatively middle-ground representations of the Troubles in terms of film and TV, are Ron Hutchinson's *Rat in the Skull* (staged in 1984 at the Royal Court, made for Central TV in 1987, re-staged in the Royal Court Classics in 1995), and Graham Reid's Trilogy *The Billy Plays* (1982–84), made by BBC Northern Ireland. Like Lynch's *The Interrogation of Ambrose Fogarty* (1982), they explore a situation ultimately rooted in colonialism, but whereas Hutchinson and Lynch are more overtly concerned with the politics of the Troubles, Reid foregrounds their effect on working class urban life. Hutchinson suggests his own work

> is also yrs. truly arguing with himself, trying to square his Northern Ireland Protestant heritage with a deeper sense of all Irishness, setting his head against his heart, trying to find a position.
>
> (1995 reprint, Foreword to play)

Lynch's play is set earlier than the others, in a police station in West Belfast during the mid-1970s, a period of unrest following the Bloody Sunday atrocity of January 1972 when British soldiers fired upon a peaceful demonstration in Derry, an event still under a Public Inquiry in Britain during 2000. Rather like *Dockers,* a contrast between ne'er-do-well Trickster-style Willie Lagan and the probably innocent Ambrose Fogarty is a crucial dramatic device. Willie, on the dole but a part-time folk singer, has been picked up during a riot, accused of throwing a beer bottle. Through this possibly accidental involvement, he is pitchforked uncomprehending into the brutal rituals of police interrogation. Married and a father, Ambrose, a Queen's University drop-out also on the dole, has long been suspected of IRA involvement by the police, who have lifted him on suspicion of bank robbery. Two policeman, who divide physical and psychological interrogation techniques between them, are determined not to believe his account of innocent activities and are enraged by his refusal to co-operate and insistence he is not involved;

> I don't believe in heroes. Life is not made up of heroes. It is made up of ordinary, unsung people trying to search out a living.
>
> (ibid., p. 157)

At first unsuccessful in bribing ebullient Willie to inform on Ambrose, the police threaten him and smash his guitar, so he pretends he saw Ambrose fire in the air. Even after a thorough beating, Ambrose insists on his innocence, throwing the statement pen away, so that finally both men are released. Evidence of local unemployment, the role of sectarianism on the job market, and the way numbers of young men join sectarian organisations, help to create the impression of an economic context where empowerment for men is difficult. Dramatically, the final phase ironically counterpoints Ambrose's attempt to file a complaint about the beating, with two of the policemen talking about their role in the community, defending their methods against those they see as hard men, as a 'contribution to peace' (ibid., pp. 161–3).

Hutchinson's disquieting play, set around ten years later, also shows interrogation of an Irish Republican prisoner, detained in Britain under the Prevention of Terrorism Act, who unlike Ambrose, seems to have been active. Commenting on the televised version, Pettit considers *Rat in the Skull* to be unusual in the way it 'is a searching revision of loyalism's core beliefs and attachments to British values' (2000, p. 239). It explores police brutality, and the passionate commitment of both sides during a period of street violence, bombing, cross-border incidents and imprisonment (including innocent individuals).

For the 1995 stage production, the theatre stalls were built over so that the London prison cell, in which the Republican prisoner Roche (played by Rufus Sewell) was interrogated by two Metropolitan Policemen and RUC police officer Nelson (played by the late Tony Doyle), was in the centre, above the audience, creating an electric atmosphere. Interrogators approached on metal walkways at cell level. Around theatre walls and balcony edge were projected pictures of actual political prisoners and other Troubles images – including considerable physical damage previously inflicted on Roche illegally by Nelson. Nelson sarcastically calls himself

> Another kind of Irishman [...] a kind of collaborator with the Army of Occupation – a traitor working for the Brits in a colonial war.
>
> (1995 reprint, pp. 13–14)

Despite his confidence, it is because Nelson assaults Roche that the case fails, and he is freed. The complex turns of the battle of words through which Nelson tries to examine 'the rat in Roche's skull' become a means of interrogating his own as well as his opponent's sectarian beliefs. These are rooted in what the nebentext calls 'the beauty at the heart of both men's obsession' – in other words, Ulster itself (ibid., p. 32). Hutchinson claims that whereas in 1984 'things were going horribly wrong', in 1995 'a play about bleakness is good in a time of generalised hope because it shows there is a chance' (*Evening Standard*, October, 1995).

Physical violence and economic difficulties – including paternal need to migrate to England to gain work – are linked to family problems in *The Billy Plays*, set in Belfast, which first brought Kenneth Branagh to a wider audience. Coulter cites research which shows that 'the contemporary Northern Irish family is a profoundly patriarchal institution' with a rigid division of domestic labour, a common male reluctance to be involved in housework or child care, even when unemployed, and – most seriously – that 'domestic violence is not aberrant, but rather systematic' (1999, p. 103, p. 108). He quotes horrific data about injuries and deaths inflicted on women, collected by McWilliams and McKiernan (1997). In *Too Late to Talk to Billy* (1982), which is structured to include poignant flashbacks, Billy's father Norman evidently experienced violence from his Da and has often, particularly when drunk, inflicted it upon his own wife and children. Inability to communicate affection prevents him from visiting his dying wife, Janet, in hospital until it is too late. Past scenes of Norman beating up Janet and her lover are juxtaposed with present-day shots of him hitting Billy, who lies prostrate, and of Janet lying ill. All the males expect domestic servicing as a matter of routine – Lorna becomes surrogate mother to her two younger sisters, Ann and Maureen. This dominant attitude to women is carried through into the sexual attitudes of Ian and John, young men whose repressed anger erupts into violence against each other – and is partly channelled into UDA-style military group involvement, offsetting powerlessness brought on by unemployment or low wages. Billy, despite seeing this swaggering as foolish, is still capable of cracking a skull and laying down protective rules for his younger sisters, as well as challenging his father, who returns to work in England.

In *A Matter of Choice for Billy* (1983), Lorna points out Billy's likeness to his father – 'It doesn't make you any less of a man to be able to tell someone how you feel' (1984, p. 91). There is some moderation of previous hyper-masculine views as their female partners persuade John and Ian to relinquish paramilitary involvement, though the latter persists in sexual infidelities as a right. This gentler tone is extended in *A Coming to Terms for Billy* (1984), when Norman returns to Belfast with his new wife Mavis, who has helped him to become more open, and after initial reluctance the two younger children join them in England. At first hostile, Ann says the Irish are hated in England, while Maureen claims, 'Anyway we're not Irish, we're Protestants' (ibid., p. 152). Ironically, Billy is only reconciled with his father when they join each other in an (offstage) fight (ibid., p. 166). Despite the personally optimistic closing tone, in the background an Orange parade bangs out the Loyalist tune 'Derry's Walls'. This play, which just pre-dates the Hillsborough Agreement, thought by many Loyalists to be a betrayal in favour of the Republicans, does not really interrogate the relationship of personal and socio-economic problems revealed to the public, political context. Reid has since written other sombre more strongly political films such as *Life After Life* (1995), about a long-term republican prisoner who finds life has moved on after his release, and *The Precious Blood* (1996), a love story between an ex-loyalist paramilitary turned evangelist preacher and a widow, who turns out to have been married to the man he previously shot.

Approximately 15 years later, in the context of the current Peace Process, more dramatically successful are the overtly political and frighteningly powerful stage plays of Gary Mitchell, a loyalist from Rathcoole, a sprawling North Belfast working-class housing estate. Interviewed by Fiachra Gibbons, he pointed out,

> There has been a fundamental crisis in Protestant culture [...] We have been going through an extremely depressing loss of identity, loss of culture, and worst of all loss of a future. The past has been hijacked by the other side. They seem to have this huge mythic Past [...]
>
> (*Guardian* 10/04/2,000)

Indeed, Pettitt's account of developments in the film and television industry both sides of the border, and especially of Neil Jordan's highly successful *Michael Collins* (1996), seems partly to support this view (2000, pp. 256ff). So does Tom MacIntyre's *Good Evening, Mr.Collins*, staged at the Abbey in 1995, a witty deconstructive piece that mocks De Valera and reveals the Don Juan side of the heroic Collins. It suggests a confident familiarity with this mythic past, combining satire with 'the man's courage, laughter, his fallible longings' ((ed.) McGuinness 1996, p. 233). From 1991 Mitchell's earlier output includes 11 radio plays, several screenplays and at least five stage plays, mostly produced in Northern Ireland. He became more widely known through *In a Little World of Our Own* (1997), premiered at the Peacock, which toured Ireland including Belfast and the Donmar Warehouse, London. *Trust* (1999), directed by Mick Gordon for the Royal Court Upstairs (at the Ambassadors), and *Marching On*, directed by Stuart Graham at the Lyric, Belfast (June 2000), deal with intergenerational conflict, suggesting 'divided families are a metaphor for a divided community' (Billington, *Guardian* 17/06/00). When younger, Mitchell's complete

Loyalism caused him to flirt with various paramilitary organisations, though not up to the necessary hilt:

> I was racked with guilt [...] People put it down to a lack of masculinity. I was full of weakness but those weaknesses I now see as my strengths.
>
> (ibid.)

His plays question not only aspects of Unionism but also the role of a tribalist attitude that ultimately destroy family, relationships and, most importantly, the truth which he sees as the first victim of the Troubles.

> When things are so obviously fucked up, no one wants to listen to the guy pointing to the styes in everyone's eyes [...] But if you're a playwright, that's your job.
>
> (ibid.)

According to Susannah Clapp (*Observer* 15/03/98), there was little initial interest in Belfast in staging *In a Little World of Our Own*, apart from the suggestion its setting should be changed to Birmingham, a point that supports Mitchell's comment. Nevertheless it was voted Best New Play at the first Irish Theatre Awards in 1997.

In a Little World of Our Own is a savage critique of ways that a hard masculinity distorts values, even within the family. Mitchell recounts how the audience 'just sat there in silence, not leaving the theatre' (ibid.). Page references below are taken from a then-unpublished script provided by Judy Friel, interviewed while she was acting as a Literary Manager at the Irish National Theatre. Events take place in the living room of a Protestant Belfast home. Unemployed Ray, in his mid-twenties, whose mother is seriously ill upstairs, is tough in every way – except for his care and concern for his disadvantaged younger brother, Richard. Another brother, Gordon, an insurance collector/salesman, is buying a house, prior to marrying Deborah, a keen Christian. Walter, over 40, is a Community Liaison Officer for Rathcoole Ulster Defence Association, an illegal Loyalist, Protestant Organisation, considered by many to be Paramilitary/Terrorist.

The complex plot involves (offstage) characters: Monroe a 'non-violent' yet influential UDA loyalist, and his fifteen-year-old daughter, Susan, reputed to be a sexual tease and leading on the innocent Richard. According to Walter, Monroe has joked he would rather his daughter 'went with a retard than a taig' (ie: a Catholic. Script p. 5). This unpleasant 'jest' increases Ray's dislike of Monroe's 'namby pamby' and 'political' approach, because he himself believes firmly in the power of UDA violence to 'sort out' local problems. 'Whether it is dealing with the IRA or dealing with petty theft or glue-sniffing' (script, p. 2), 'A few broken bones and bruising will be forgotten about in a couple of months' (ibid., p. 3). Richard, supported by Ray, is reluctant to consider living with Gordon and Deborah in the future, a source of tension between them. When Ray drops Richard off at a party where Susan will be, it transpires the girl goes missing, and Walter implies Richard is widely suspected of involvement. When Ray and Richard return, it at first seems that, after being disturbed by the brothers, the taig has escaped having beaten and raped Susan, who later dies in hospital. It is not clear to the family and Walter whether Richard – obviously distressed and having been drilled in his evidence by Ray – was in fact the guilty party. Masculine tribal loyalty underlies the need to cover-up through finding a (Catholic) scapegoat. Insisting the taig was responsible, Ray leaves to set up a video taped interrogation of the 'suspect' in a flat when he finds him. Richard indavertently reveals to Walter that Ray had 'sorted it' by raping and beating the girl, the audience presumes in revenge both for her mocking behaviour in leading Richard astray and for her sexual involvement across the sectarian divide. This tense process of revelation culminates in Gordon's need to 'kneecap' Richard as a punishment likely to be approved by the UDA;

> They're honourable men and they are giving me the opportunity to do it first. Like a family thing. But if I don't do it then they're going to come after us [...]
>
> (Script p. 97)

Details of the matter could thus be withheld from the official police force (the RUC), so Ray, a significant member of the tribe, would not be suspected. Religious imagery of sacrifice and scapegoating resounds through their dilemma. Ray returns shot through the stomach – 'things have gotten out of hand' (ibid.) – wanting to see his mother, with the news of Susan's death, saying Walter has betrayed him to the UDA. Gently telling Richard the girl is dead and that he himself will 'have to go and look after her' (script, p. 105), Ray sends him upstairs. When Gordon cannot bring himself to execute Ray, Richard suddenly appears, shoots Ray and hugs him as he dies. The virtual absence of the mother figure, and rather unappealing representation of Deborah, emphasise masculine values. The cumulative force of revelation and counter-revelation in performance avoids potential melodrama and brilliantly conveys the writer's anger about the corrosive effect of this kind of tribal violence, which turns sour essentially good qualities such as loyalty, community solidarity, religious morality and support.

Linking masculinity, violence and unemployment with sectarianism, *Trust,* first performed at the Royal Court Upstairs (the Ambassador's) in March 1999, has a similar claustrophobic feeling: set in a Belfast house, there is one brief pub and one exterior scene suggested by lighting change rather than furniture removal. The play's two women, Margaret mother to fifteen-year-old Jake, and Julie, Vincent's girlfriend, are in different ways quite strong, but misguided. Margaret is protective of her rather studious son, while his father Geordie with drinking partner Artty feel they need to make a man of him. Despite Jake's reluctance, they take him drinking and mock his interest in homework and computers. Although Jake (just) passed qualifying examinations for a good school, Geordie insisted Jake attended a local one where he can play football with his peers, but he is being bullied, and unable to fulfil his academic potential. Margaret mocks Geordie's attitude – 'What would everyone think if my son went to a posh school?' (1999, p. 55). Like Artty, Geordie believes the best aid to learning and crime avoidance is 'a quare slap around the lugs' (1999, p. 6). Though unemployed, Geordie's important local status as UDA hard man is shown when Trevor, recently released from thirteen years in jail, comes round to seek help. Geordie asks if his Ma is now alright,

> If there is any more bother she's just to send you round and it'll be sorted again. Maybe permanently if she wants like.
>
> (1999, p.20)

Margaret's frustration that Geordie will not help his own son has serious repercussions when she enrols Trevor to help 'sort out' the bullying boys from the Turkington family. Trevor reports,

> I told him to hit the wee lad but he wouldn't [...] I had to give the wee lad a bit of a hammering, but even still Jake wouldn't get involve.
>
> (1999, p. 57)

He has given Jake a knife, which, panicking he later uses to stab one of these boys. Geordie and Artty continue macho boasts, that once the fuss dies down the boy ' [...] will be going back into hospital and so will his brothers' (1999, p. 67). A parallel plot involves Julie's attempt to get guns for the UDA through Vincent, her soldier boyfriend, which culminates in their arrest after the UDA have received the weapons. Geordie and Margaret remain entrenched in their different views – he would prefer to deal with the injured boy's family. Pulling one of the (unloaded) guns on her husband, she is determined to negotiate with the police, to trade her son's safety from prison for information about the guns' whereabouts. The play suggests the corrosive effect of misplaced 'trust' between hard men, whose tribal notion of masculinity distorts relationships both across the generations, and between the sexes, where women are also drawn into violence.

Intergenerational violence and faulty communication on both sides of the Border are thus relevant to the fourth theme: the father's identity and familial role. Meaney's application of Nandy's postcolonial theories about India to Ireland is relevant:

> Anxiety about one's fitness for a (masculine) role of authority, deriving from a history of defeat or helplessness, is assuaged by an assumption of sexual dominance.
>
> ((ed.) Smyth, 1993, p. 233)

Patriarchal violence may thus be a displacement activity, passed from one kind of victim on to another. Edna O'Brien's *Our Father*, directed by Lynne Parker at London's Almeida Theatre in 1999, used a realist form and fourth-wall setting to convey the effects of patriarchal power on daughters in rural life during the 1970s, through showing different siblings' reaction to the experience of their repressive father's funeral. In comparison with other current plays, O'Brien's treatment of this theme seemed bland and old-fashioned. Carr's *On Raftery's Hill* (2000), directed by Garry Hynes, was a Druid/Royal Court joint production. A piece that combines black humour and tragedy, it was rescued from melodrama both by the intensity of the actors – especially Tom Hickey as Red Raftery – and the pitying irony and poetic skill of Carr. This treatment of incest – like the sexual abuse of children by priests, the treatment of orphans by nuns, or the babies found buried in a Kerry field – is part of the new openness in Irish society toward issues, which, though true, are difficult to believe, as acknowledged by Hynes in interviews with Lyn Gardner (*The Guardian* 3 April 1996, 3 July 2000). The kitchen set, with twisted staircase upstage and greenish decaying walls with a touch of dung brown, echoed the offstage decay of Raftery's farm, with its potentially beautiful land full of rotting carcasses of sheep and cows. Patriarchal violence seemed linked both to self-destructive qualities and female compliance as increasingly horrific details tinged with black comedy emerge. Red's mentally and emotionally damaged son, Ded, lives in the cowshed like an animal – banished, it later transpires, because he had taken Dinah, the elder daughter to the shed to give birth as a result of long-term incest with her father. Hints imply that on Red's orders, Ded, once his dead mother's favourite, had 'cleaned up the mess', that is, killed the baby. Dinah, sent by their mother into Red's bed when only twelve years old, has tried to protect her younger sister, Sorrel, from their father. Red's perverse rages are also about keeping his land sacrosant: not believing in forgiveness, he feels he and his now demented mother Shalome are

> big loose monsters [...] no stupid laws houldin us down or back or in.
>
> (2000, p. 30 & p. 42)

Neither does he disapprove of an incestuous neighbour, who drinks weedkiller after the death of his daughter and their baby come to light. Shalome, hag-like in her dementia, constantly tries to return to her long dead, brutal father, whom she seems to have loved inordinately – while cruelly keeping Old Raftery her husband at bay. Remembering her childhood in India, she taunts Red that he is not the son she wanted, but laments she was forbidden to send him away to be educated:

> Old Raftery wanted you rough and ignorant like himself.
>
> (ibid., p. 25)

Act One ends with Red's brutal rape of Sorrel, whom he heard discussing both his rages and possible ownership of some family land with her fiance, Dara, who describes his own father as 'one sad picnic in the rain' (p. 32). Another grotesque male, widower Isaac, obsessed by his aged cat, prefers animals to people. Act Two shows how Sorrel, convinced that 'daddy [...] just wants to bate us all inta the dirt' (p. 37), is unwilling to co-operate with Dinah's attempts to carry through the wedding, funded lavishly by Red. When Dara proudly refuses to take some land and a large cheque from Red as a wedding present, Sorrel refuses to marry him – as Shalome escapes again, symbolically dirtying the wedding dress. Although the revelation of such rural horror is socially constructive in reality, in fiction Dinah's compliance and Sorrel's defeat are hard for feminists to accept. The play shows Dinah's positive memories: 'He knew how to build up a child's heart [...] Daddy' (p. 40), while even Sorrel claims to Dara, 'He's good at the back of ud all' (p. 54). Dinah's statement 'We're a respectable family, we love wan another' (p. 58) has a tragic subtext that echoes the disjunction between expression and experience, which has been traced throughout this chapter's analysis of masculinity. The persistence of patriarchal power is stressed in Salome's last speech:

> Daddies never die, they just fake rigor mortis and all the time they're throwing tantrums in the coffin, claw marks on the lid.
>
> (p. 59)

The absence of the mother – dead after giving birth to Sorrel in McPherson's play – seems to be a crucial factor in giving further space for male violence. Gina Moxley's *Doghouse,* set in a cul de sac in Cork city, was commissioned by the annual Connections project for new plays written for young adults, work-shopped by the National Theatre Youth Company in London and published (1997). Serious offstage events are perceived only through the eyes of young people who live there, amidst their other more typical concerns. Not only has the mother of one family temporarily withdrawn through nervous trouble, but the mother of the impoverished Martin family, who move into the street, is also dead. Moxley effectively conveys the horror of brutal Mr Martin, who is never seen. His drunken cruelty – and cunning in using his elder daughter, Val, as an agent of his power and another daughter Bridgit as a messenger, becomes evident in the damage he inflicts on Pats, his child scapegoat, and his son Dessie who tries to support her. When young, but slightly older, neighbours begin to realise that not only is Pats starving but also physically abused, they send for the police in time to prevent Dessie taking a shot at his father. Mr. Martin, who cares more for his dog than his children, is fortuitously killed in a car accident when he swerves to avoid a dog – ironically not his own. The last scene shockingly exploits the disjunction between the cultural myth of an ideal Irish family life and the harsh reality for some, when during the final funeral scene Pats flings the earth into the grave. She laughs hysterically when a dog is heard barking offstage, infecting all the other young mourners.

McPherson's *Dublin* Carol, directed by Ian Rickson for the Royal Court's re-opening in January 2000 also provides an urban context for another though more positive example of dysfunctional father/daughter relationships. The play is set in a funeral director's grimy office with a few tatty scraps of festive decoration and limited kitchen facilities. On Christmas Eve, John Plunkett, an undertaker in his fifties who had lost contact with his family, is advising Mark, his young assistant. Mark's Uncle Noel, the boss, who is now very ill in hospital, had befriended John when he was in a terrible, rock-bottom mess. A gentle sense of male solidarity is implied in these relationships. Part Two shows John talking to his estranged daughter Mary, played movingly by Bronagh Gallagher. After ten years Mary has come both to ask him to visit his dying wife in hospital, and to give news of his son, Paul, who works in England. This confrontation causes John, who now drinks less, to review his past when he abandoned wife and family due to involvement with Carol, a mistress who supported his drinking. Expressing feelings of social alienation, Mary claims to be like her father; while he tries to explain his feelings of cowardice and inadequacy, originally triggered by the way his own father beat his mother and himself. Part Three shows John waiting to be collected by Mary, talking to Mark, revealing his wish for a ' … clear signal … to people I mean' (2000, p. 51). Finally after washing and brushing up, he restores the decorations he has previously taken down, placing a star signifying hope at the top of the small Christmas tree. Forgiveness and willingness to change are shown as positive masculine possibilities, not weakness.

Within dysfunctional families explored in this chapter, sons have also suffered different kinds of damage, often through the patriarch, ranging from a reluctance to educate sons, to Ded's extreme breakdown, to the migration of abandoned Paul, or Dessie's attempt to kill his father. On a gentler level, the sons of the comparatively successful long marriage of Dan and Doris manifest other weaknesses in Bernard Farrell's ESB/*Irish Times* Award-winning *Kevin's Bed* (1998). Blending comedy and compassion, it explores family relationships by fading in and out of the times of Silver and Golden Wedding Anniversary celebrations of Kevin's parents. Both parties are suggested by offstage popular music, and noise. Farrell, who has written plays for the Abbey, Red Kettle Theatre Company and award-winning *Stella by Starlight* (1996) for the Gate, was elected to Aosdana in 1992. Although the use of time is formally disruptive, the play's overall tone is predominantly nostalgic rather than socially critical in its exploration of unfulfilled dreams, with humour and plot twists relying upon the device of a kitchen 'dumb waiter' as an ineffective means of communication. Crucially, sibling rivalry exists between Kevin and John, who married and later divorced Betty – who loved Kevin. In both past and present, ineffectual Kevin lacks courage to reciprocate openly. John, once a successful schoolteacher, went to England and fell to pieces after being imprisoned after a sex scandal. Kevin spoiled the Silver Wedding by abandoning the priesthood and marrying a determined Italian girl, who finally calls him 'a bad husband, a bad father, and a bad friend and a bad lover' (1997, p. 106). Eventually left alone at the Golden Wedding, he remembers his unfulfilled past, ignored even by Betty. In contrast, Dan, a dominant patriarch, fulfils his wife's dreams by regaining possession of this house – where Kevin and family have been living for years – through lying that Doris is fatally ill. Farrell's popularity and skilful touch perhaps shares some of the craftsmanship of Hugh Leonard, whose 'light' plays generally tend to evade 'social or political impulse' and have thus not been discussed here. (See Murray, *Irish University Review*, Spring, 1988.)

The fifth theme for discussion in this chapter is evident in two plays which differently explore gay issues; Kilroy's *The Fall of Constance Wilde* (1997), and Stembridge's *The Gay Detective* (1996). The Irish Gay Rights movement was founded in 1974, but legalisation of homosexuality was not achieved until 1993. According to Cullingford,

> the emergence of overt representations of gay men coincides historically with the reform of Irish anti-homosexual legislation that was initiated by Senator David Norris's 1988 victory in the European Court of Human Rights and completed in 1993.
>
> (in (ed.) Bradley & Valuilis, 1997, p. 176)

Her interesting exploration of representation of gay men and of the role of homoeroticism in the postcolonial context; like Frazier's essay on 'Queering the Irish Renaissance' (ibid., pp. 8–38) or others in *Ireland in Proximity* ((eds) Brewster et al., 1999) show that this is too wide a topic to be covered fully here, especially in lieu of publication of recent relevant plays. Walshe's essay 'Wild(e) Ireland' in Brewster's collection explores the emergence of a gay culture, partly through comparative analysis of representations of Oscar Wilde in plays.

Whereas Eagleton's *Saint Oscar* (1989), staged by Field Day, and seen on British television, makes an explicit link between potential subversion of colonial power through dissident sexuality, deploying a disruptive approach to time, memory and the discourses of Wilde's circle, Kilroy's mediational modernist play focuses more upon the construction of sexual identity per se. Through the perspective of Wilde's wife, Constance, it is suggested – as in Eagleton's play – that his desire was for the polymorphous pleasure of the androgyne – 'I must have it! I will have it! Neither man nor woman but both' (1997, p. 20). Seen at the Abbey in 1997, though also briefly transferred to London's Barbican in autumn 2000, its strong visual power was indicative of Patrick Mason's direction. Detailed nebentext suggests the stylised movements of mysterious white-masked, black-clad attendant figures who constantly watch, clap, perform tableaux and facilitate the occasional use of puppets to represent the Wilde children. Other visual details include a white disc within which three protagonists – Oscar, Douglas and Constance – move, a circular stairs and a cage. Events occur in a disrupted time sequence, and when two characters are in intimate dialogue, the third always watches from darker parts of the stage, providing a further inner audience.

From the outset, Constance is aware of the disjunction between experience and expression crucial to Oscar's nature:

> You never face the situation as it really is. Never! Nothing exists for you unless it can be turned into a phrase.
>
> (1997, pp. 11–2)

This 'gap between consciousness and action' and 'the familiar Irish discrepancy between rhetoric and reality' is linked by Eagleton to Wilde 'adopting a performative rather than a representational epistemology' (1995, p. 333). This play suggests Constance, who feels she has been 'invented as a good woman' (1997, p. 13), was drawn to Oscar partly as a result of her monstrous secret only revealed near the end – that her father, who was arrested for exposing himself in public, also sexually abused her. Her fall is both literal (down stairs) and metaphorical. Although Constance is shown to be unusual in her understanding of Oscar's sexuality, and courageous and angry in her hate of hypocrisy, the play contextualises homosexuality in suffering, trauma and disgrace. Douglas' love of pleasure and money is shown to be stronger than any genuine feeling for Oscar – while both seem to have endured repressive, unpleasant fathers. Theatrically effective in both structure and devices – such as a gigantic incestuous father, a Black Mass and the Androgyne image repeated as a transformation of Oscar at the end – its representation of homosexuality is rather ambivalent. However, the future possibility of more flexible, blurred sexual identities is evoked both by Oscar (1997, p. 52), and Douglas (ibid., p. 56), whilst Constance's last letter to her sons defends Oscar as a great man who sacrificed all for his vision.

Where Kilroy's play teems with intertextual references to Wilde's plays, Stembridge draws strongly upon the traditions of film noir and crime narratives. Born in Limerick,

Stembridge, a successful director, has written eleven stage plays and written for radio and TV, with an award-winning debut film, *Guiltrip* in 1995. Written in 1996 about 1993, *The Gay Detective* refers to the (then) present illegality of gay activities, and to imminent likely change in the law in Ireland (1996, p. 32). It also makes tongue-in-cheek jokes about the topicality of the issue – as likely to feature in Gay Byrne's TV show with a media psychologist and to prompt a book by Fintan O'Toole (1996, p. 41). Shown originally at Project Arts Centre, Dublin, it features several openly erotic promiscuous gay encounters, yet is ultimately a love story between Pat, the Gay Detective, and Ginger, whom he encounters when tracking down a 'queer-basher'. Apart from these two roles, the others are all doubled, and given animal names. The story, introduced by Pat, has comic touches. The nebentext suggests fluidity of form and changing of basic costume/roles, preferably with all actors remaining on stage somewhat in Brechtian style. On realising the efficient Pat's sexual orientation, his Sergeant Bear places him undercover in gay venues, in the hope of tracking down a drugs baron. Events become increasingly more complex and violent once a Government member is found murdered. Social hypocrisy is foregrounded in the composition of a group of powerful closeted homosexuals – a top criminal lawyer, a famous classical musician, and the Government minister (before his murder). Wolf, a prominent businessman, possibly a drug baron, helps the group exploit others – especially working-class 'rent' boys – leading to the disappearance and death of one, and almost the death of undercover Pat. The Sergeant, however, is not prepared 'to arrest three distinguished citizens for the murder of a little shit, a pansy rent boy' (1996, p. 76), when Pat reveals the sado-masochistic circle instead of the drug baron. Bear is only interested in arresting them if they had killed the government official. Having endured a near-fatal encounter with Wolf, Pat helps his exploited servant, Mouse the killer, to leave the country and returns to comfort Ginger who is suffering from HIV. The play's moral perspectives are as complex and confusing as the overall plot. In contrast to casual fulfilment of Pat's sexual desires, some in the cause/course of duty, there are increasingly tender moments with Ginger – temporarily suspended when Pat first hears about the HIV – and concern from Puppy, a girl neighbour. This formally radical modernist play gives the opportunity for homophobic statements to be aired, especially by the Sergeant, in contexts where they are by implication open to critical reading by the audience. Despite Pat's sense that he has betrayed the gay community by working undercover for the police, he is finally able to confess his love to Ginger. The film-noir style perhaps makes it more difficult for fully positive representation of gay men's experience to be realised on stage. Whether 'the 1990s reforms have delivered gays from insidious homophobia' is still an open question (Pettit, 2000, p. 273).

In overview, this chapter seems to present a rather bleak view of the representation of masculinity both in the Republic and in Northern Ireland. Apart from a more poetic perspective on the harshness of rural poverty in *Boss Grady's Boys* – plays analysed suggest that social and economic factors arising from the colonial past, produce a tension between expression and experience that tends towards patriarchially sanctioned physical and emotional violence in the countryside. This seems almost inevitably worse in an urban context, especially when exacerbated by sectarianism. However authentic the setting, the predominance of the ideologically restrictive realist form may be responsible for a more fixed notion of gender, whereas those plays with mediational modernist form tend to approach the question of identity in a more fluid, interrogative way. Representations of the cultural 'Dream of the West' and the diaspora, both responses to those frustrations bred from political and socio-economic difficulties before the Celtic Tiger and the Peace Process, will be explored in Chapter Six. The more flexible notions of gender identity/identities introduced during this chapter will be discussed further in Chapter Seven, with further reference to open-ness of dramatic form.

6 Dreams & Diaspora: Exported Images and Multiple Identities

This chapter explores ways in which the Irish experience of migration is represented in contemporary drama. The relationship between Ireland and America has often been associated with 'a Dream of the West' on both sides of the Atlantic, whereas representations of the Irish in Britain may reflect or refract the proliferation of derogatory stereotypes. Two quotations, cited in Lance Pettit's *Screening Ireland* (2000), indicate the importance of images of Ireland to the diasporic community: 'emigration is at the centre of the Irish experience of being modern' (in (ed.) Kearney 1990, p. 47), and 'the emigrant's break with the past has been internalised within Irish culture, forming popular images of itself' (Gibbons, 1996, p. 40). These two ideas are germane to the paradoxical relationship between images of Ireland and the Irish, which promotes identities readable as either conservative, simplistically unified in their subjectivity, or radical and complex in their multiplicity. Although both the post-colonial context and the diaspora have a significant role in creating a market/audience for such images, further ideological implications underlie these different approaches to cultural identity.

A key figure in Irish drama is either the individual who decides on exile, or the family member whose return visit from abroad acts as a catalyst, challenging the identities and lifestyles of those who have stayed at home. Within each thematic chapter, at least two plays have contained such figures: for example, in Chapter Two, Murphy's *Famine* (1968), Friels's *Translations* (1980) and *Dancing at Lughnasa* (1990); in Chapter Three, McGuinness' *Mutabilitie* (1998), *Mary and Lizzie* (1989), and Devlin's *After Easter* (1994); in Chapter Four, C. Reid's *The Belle of Belfast City* (1989) and McDonagh's *The Beauty Queen of Leenane* (1997); in Chapter Five, Roche's *Poor Beast in the Rain* (1991), G. Reid's *The Billy Plays* (1982–3) and Bolger's *A Lament for Arthur Cleary* (1989). All provide examples of migration and/or return. This chapter explores several issues. First, a brief exposition of contextual issues with framing reference to Friel's *Philadelphia Here I Come* (1964) and Declan Croghan's *Paddy Irishman, Paddy Englishman, Paddy ...?* (1999), indicates American and English aspects of the diaspora, but with relevance to the decision to leave Ireland. Second, comparative analysis of McDonagh's *The Cripple of Inismaan* (1996) and Jones' *Stones in His Pockets* (1999) explores the relationship between the process of filming – that is manufacturing of exported images of Ireland – and local socio-economic conditions likely to be relevant to migration. The authenticity of such images is also considered. Third, brief analysis of three plays whose style and content reflect aspects of the American dream is followed by reference to seven recent plays which explore the role of the returning migrant – whether temporary or permanent – including two about problematic wakes. Finally, the downbeat approach to such returns provides a stark contrast with images of Ireland apparently intrinsic both to the tourist industry and popular television comedy programmes shown in Britain.

Cormac O'Grada's account of the Great Famine indicates significant migration figures:

> Between Waterloo and the famine 1.5 million left Ireland for good [...]
>
> (1989, p.8)

> in 1846 a 100,000, and for five years after 1847 200,000 left annually, falling to 70,000 annually by 1855.
>
> (ibid., pp. 47–8)

Most of these migrants went to the USA. In following years there has been, until the relatively recently economic boom, a steady haemorrhage of the Republic's population to different destinations, according to the economic state of Britain, America and Europe relative to that of Ireland. According to O'Toole:

> One in every 12 people living in the Republic in 1982 (ie. 289,000 people) had emigrated by 1989 when outflows were temporarily staunched by recessions in Britain and the United States.
>
> (1994, p. 12)

He also quotes a survey of families in Cork city, South Limerick, South Kerry, East Galway and North Donegal. This revealed that a quarter of these 'had at least one family member living abroad' with one in three of the Donegal families having a recent migrant, and one in ten having four or more (ibid., p. 13). O'Toole also refers to American President John F. Kennedy's visit in the 1960s to his family's Irish roots, when he referred to people as Ireland's only export, while heralding potential economic re-investment from countries enriched by this migration back in the home country. According to Lyons, emigration rates for 1951–1961 were 'higher than for any comparable period in the twentieth century' (1973, cited Smyth, 1998, p. 94), mostly for English destinations. Especially in the eighteenth and nineteenth century, migration also took place from what became Northern Ireland as, for example, shown by the Ulster-American Folk Park built near Omagh to celebrate thousands of Ulster-Protestants who left for America. David Brett's critique of this park's 'staged authenticity' nevertheless acknowledges that 'the intellectual lineage of Dissent leads directly towards the Constitution' (1996, p. 118). Whereas in the past, as described in Liam O'Flaherty's heart-rending story *Going into Exile*, a family may have held an 'American Wake' prior to a loved one's departure – because it was unlikely they would ever be together again – changes in economy and technology have now made communication and return travel easily accessible, with the potential for flexible working within a global economy. Where 60,000 people left in 1988, 42,000 entered Ireland in 1999 (*The Guardian*, 21/09/1999). Indeed, whereas original emigrants more often had only basic skills, now it is young professionals who move in both directions. In terms of cultural identity, as Kearney points out, 'If over seventy million people in the world today claim to be of Irish descent [...]' any definition of nationality 'extends far beyond the borders of a state or territory' (1997, p. 9).

Politically, not only was the original development of Irish republicanism in the 1790s influenced by the American Declaration of Independence but, according to Kearney it was 'American in practice (but) largely French in theory' (ibid., p. 53). American money has supported Fenian campaigns, and in the 1940s De Valera visited America. American-Irish involvement in the Northern Irish Peace Process has been epitomised during December 2000 in retiring President Bill Clinton's second visit to both the Republic and the North, to reinforce his support for finding a lasting solution. Welcomed in Dublin by Prime Minister Bertie Ahern and President Mary McAleese, Clinton chose to give his keynote speech at Dundalk, a town which is not only an example of new Ireland's industrial growth but also tainted by the possibility it had harboured members of the Real IRA, responsible for the bomb atrocity that had killed 29 people at Omagh across the Border. Clinton then went on to Belfast.

The relationship between Ireland and America is thus long and paradoxical, encompassing tension between economic hopes and emotional anxieties which are reflected in drama's representations, both of the migrant's aspirational dreams and of diasporic dreams of home. As a concept, the West has long been associated with heroism, mystery and romance: in modern times, these notions have been further fed by the American entertainment industry, especially Hollywood film. Where O'Toole feels 'the myth of America as a refuge is something that has paralysed us for a very long time' (1994, p. 77), Gibbons' analysis of 'the Dream of the West' compares the Literary Revivalist dream of the West of Ireland with the Irish dream of America as an ideal place (1996, pp. 23–36). He suggests that while both evoke an agrarian ideal and a hostility to centralisation as well as the power of the law, the West of the Revivalists was valued for its conservative communitarianism in sharp contrast to the American West's emphasis on individualism. In juxtaposing masculine archetypes of playboy and cowboy, Gibbons suggests the former embodies possibilities of consumption and pleasure, whereas the latter can be, like the puritan ethic, linked to a belief that with hard work there is a new world to be won. In this sense, with reference to Alan Price's analysis of Synge's play, Gibbons suggests the American community needed the heroic individual, whereas the Irish individual needed the community. Price observes that while Synge's playboy finally 'goes significantly developed and enriched [...] no opportunity is similarly given to transform the people and their surroundings, and the problem remains' (qtd Gibbons, 1996, p. 31).

This dichotomy between playboy and cowboy is evident in motivations for migration and other attitudes represented by dramatic characters discussed during this chapter, while the problem of stasis in the family and community left behind is especially foregrounded where an individual returns home. Plays analysed include many characteristics of postcolonial drama as previously indicated. Further, Kearney's terms – revivalist modernism, mediational modernism, and radical modernism – which indicate the nature of the closeness or distance of their relationship with traditional concepts, are relevant to ways in which images of Ireland may be created and exported in the diasporic context. The notion of simplistic stereotypical identity is, of course, challenged by the existence of multiple cultural identities within the diasporic population.

Two plays, one about America and one about England, provide an introduction to the role of images of Ireland in the diasporic context. Brian Friel's *Philadelphia, Here I Come*

(1964) has been labelled by Richard Pine (1990, p. 1) as the start of contemporary Irish drama. Its mediational modernist technique splits protagonist Gar O'Donnell into Private Gar and Public Gar, played by two different actors. This split identity epitomises the dilemma of the emigrant, who for personal and economic reasons does want to go to America, and yet fears in leaving his home community he will lose some intrinsic aspect of his identity and culture in the impermanence and anonymity he partly desires. In Gar's final encounters with friends and family, he tries to convince himself he hates his hometown, Ballybeg, yet at the same time he struggles ineffectively to establish the validity of his memories, especially to communicate a definitive image of rowing on the Lough.

> [...] once upon a time a boy and his father sat in a blue boat on a lake on an afternoon in May, and on that afternoon a great beauty happened, a great beauty that has haunted the boy ever since, because he wonders now did it really take place or did he imagine it.
>
> (1984, p. 89)

His father's inability to recall this event intensifies Gar's inability to be sure about why he is leaving. Thus this kind of contradiction between staying and leaving sets up the need for both those at home and those in exile to create images of their different contexts which assuage anxieties on either side of the ocean: to evoke possibilities both of a place to return, and of a place of success elsewhere. Pettit (2000) and Gibbon (1996), in their respective analyses of certain popular films and photographic postcards, discuss images of landscape and character whose underlying aspects of nostalgia and melancholy may resonate with the diasporic population.

In contrast to such consolatory images, Declan Croghan's *Paddy Irishman, Paddy Englishman, Paddy ...?* (1999), a black comedy, re-works Sean O'Casey's tragedy *Shadow of a Gunman* (1923). Croghan's early plays were performed at the Eblana in Dublin. This production originating from Birmingham Repertory Theatre at The Door, was shown in London's Tricycle, Kilburn, a venue strongly involved with the local Irish community. It reveals unattractive aspects of exile – the life of two young male labourers in a grimy urban bedsitter, one of whom, Anto, has drunkenly boasted he is a secret Republican activist. In this instance, the young West Belfast girl from the room above not only takes charge of the possibly bomb-containing case left with them by a real dissident gunman, but turns out to be still actively involved despite the start of Peace Process negotiations. At home in the past, she had seduced British soldiers to their execution, because her brother had been killed. Both young men leave, Kevin the Kerry man to take his electrical trade to Africa with the McAlpine company. This play's relevance to the overall discussion lies not only in anti-Irish racism the lads have experienced but, more significantly, in the economic reasons why they left home and their sense that life there was stifling – although overseas, they need to hide their emotional vulnerability in the pub craic (1999, p. 39). In spite of difficulties, Kevin still feels migration is his only chance to bloom,

> To become a full human being, reach out, touch the world, live learn and think freely, have new ideas, my own thoughts for myself.
>
> (ibid., p. 83)

Anto, from inner city Dublin, the least skilled man, with no clear idea of his own future aims or why he left, always feels he has to make a financial splash for the family when he returns home, to give an impression he is doing well in Britain. In reality both men still socialise almost entirely with other expatriates and feel victimised at work. Neither their images of home – despite their jokey references to Kavanagh's classic novel *Tarry Flynn* (1948) and to Nationalist and other folk songs – nor their present ambivalent status are comforting. Croghan implies this is a transitional time for Irish identity through Anto's closing statement to Una:

> The war is over [...] You've been ordered to stop, stop and live.
>
> (ibid., p. 101)

The play's title implies a movement away from stereotypical naming of Irish Otherness, and its reworking of a canonical text not only reveals the underbelly of the migrant's dream of elsewhere, but also takes further O'Casey's questioning of violent heroics. In terms of the playboy/cowboy dichotomy, Anto embodies some of the former's pleasure-loving characteristics, while Kevin's determination to work further afield is more typical of the individualist pioneer.

Both *Philadelphia, Here I Come* and *Paddy Irishman, Paddy Englishman, Paddy ...?* support Pine's notion that there is a gap between private, inner questioning and the ambivalent qualities of the public world; one which leaves the identity of the exile positioned in a liminal state (1990, p. 82). This borderline terminology resonates with Foucault's idea that theatre may be a heterotopic space, a site in which

> all the other real sites that can be found in a culture, are simultaneously represented, contested and inverted [...]
>
> (1986, pp. 22–7)

In other words, dramatic – and filmic – images are likely to a greater or lesser extent to encode oscillating and ambivalent attitudes to home endemic in a diasporic context. The two plays discussed in detail next, indicate ideological differences entailed by conservative or deconstructive approaches to such representations of home – and their contemporary function in terms of marketing culture for tourism.

McDonagh's *The Cripple of Inismaan*, directed by Nicholas Hytner, was premiered in London at the National Theatre's Cottesloe in December 1996, when he was Resident Playwright. Jones' *Stones in His Pockets*, originally performed in the Lyric Theatre Belfast in June 1999, transferred to London, first to the Tricycle, then the New Ambassadors in May 2000, then the Duke of York's – and thereafter to Broadway. Both plays are directly concerned with the film industry's construction of images of Irish identity and incorporate significant references to the making of actual films: in the former case, to Irish-American director Robert Flaherty's famous documentary *Man of Aran* (1934), and in the second to John Ford's classic *The Quiet Man* (1952), about the return of an American emigre to Ireland. The making of both these films is therefore affected by the diasporic context in a way which raises problems of authenticity. *Man of Aran*, with its archetypal struggle of men versus the elements, is indicated by Pettit to be flawed in its reconstructive methods (2000, p.78ff), while he links *The Quiet Man*'s excess and visual pleasures to the moment when the need for reassuring popular images of return was created by the emigrant's break with the past (ibid., p. 66). Both films were made before present incentives, such as the Republic's favourable tax legislation in 1987, further extended in 1996, the relaunch of the Irish Film Board in 1993, and developments in the North where independent companies have worked with the BBC and Channel 4. These elements have not only fed a boom on both sides of the border, with film-makers from elsewhere flocking to make films in Ireland, but have also fostered indigenous films whose take on the contemporary situation is, as Pettit shows, very different from the films discussed here.

Pettit suggests that, ironically, both *Man of Aran* and *The Quiet Man* helped to

> [...] accelerate the contamination of the culture for which both Irish Americans professed so much admiration.
>
> (2000 p. 79)

Both plays address this issue of contamination, but *The Cripple of Inismaan*, for a variety of reasons, is less satisfactory as a critique of the image-making process. The writer, despite the recent return of his Irish parents to Ireland, seems to be looking back at a community from outside. A black comedy, it draws satirically upon the outsider's 'Dream of Ireland' and the Irish 'Dream of the West' as reciprocally feeding upon each other. It echoes indigenous melodrama tradition in various details, such as the death of the eponymous hero's parents and his incipiently fatal TB, but its status as a postcolonial drama is suspect. Events centre upon a small village on the island of Inismaan, where bored orphan Cripple Billy is looked after by his two shopkeeping aunts Kate and Eileen and teased by mischievous Helen and her dim brother Bartley. Hearing from local gossip Johnnypateenmike that Flaherty is filming *Man of Aran* on Inismore, the adjacent island, Billy overcomes his fear of water. By using a forged doctor's letter stating he has a fatal illness, Billy persuades Bobbybabby, a fisherman, to let him join Helen and others in crossing to Inismore in the hope of getting into the film. Scene Five ends the first half of the performance with anxious aunties receiving news that Billy has been taken to Hollywood on the promise of a screen test for a different film. Scene Six shows the resumed boredom of village life six months later, but Scene Seven shows Billy, apparently ill in a squalid Hollywood hotel room, soliloquising in an excessively melodramatic way – in rhythmic quasi-poetic language that draws upon cliches, with the song 'The Croppy Boy' interspersed with his coughing. Imagining the death-bringing banshee's approach, he calls himself,

> Just an Irishman. With a decent heart on him and a decent head, and a decent spirit not broken by a century's hunger and a lifetime's oppression.
>
> (1997, p. 53)

He then gazes at a woman's photograph, asking his 'Mother' whether heaven is even more beautiful than Ireland. Only in Scene Eight, when Billy appears back at home as the local screening of *Man of Aran* finishes, does he reveal to friends and family that he spent hours learning 'the arse-face lines they had me reading for them in a hotel room'. He then quotes some of the above lines from Scene Seven, thus deconstructively exposing them as fiction (ibid., p. 63, p. 68). The ambivalent 'truth' of this tale makes it hard for the audience to believe Billy's tale that he has come home, rejecting the offer of a part because he cannot bear to be away from all in Inismaan. Bobbybabby beats him with lead piping for his lies. Scene Nine reveals Billy now does have a fatal dose of TB. Further, the tale his parents drowned themselves so that insurance money could help their crippled son is revealed (to the audience but not to Billy) to be yet another of Johnnypateenmike's stories. After rescuing Billy, whose parents tried to drown him in a stone-filled sack, Johnypateenmike had used his own money to pay for Billy's medical needs. Ironically Billy too takes up a sack for stones to drown himself, but is temporarily distracted by Helen's offer of future kisses. In the final moments, he is seen to cough up blood.

At the Cottesloe, the interior set of the shop/village hall space, where domestic action occurs, was dwarfed by a huge black and white photographed backdrop of sharp cliffs and bare terrain, familiar to anyone who has seen the Flaherty film. As this inner set realistically conveyed sparse, everyday details of mundane village life, it deliberately jarred with the outer frame's 'big screen' version of the Aran landscape's filmic grandeur. Dialogue throughout suggested the element of 'fabrication' in film-making. Billy is vexed that, although he was a real cripple,

> The Yank said, Ah, better to get a normal fella who can act crippled than a crippled fecker than canna act at all.
>
> (ibid., p. 66)

Helen is angry the film is not called *The Lass of Aran* 'Not some oul shite about thick fellas fecking fishing' (p. 51), and apart from their disbelief about a shark's presence, the villagers virtually ignore the film when it is screened for them. Further, conflicting discourses around the truth about Billy's parents, like the mystery of what really happened to Billy in Hollywood, erode the images of both 'home' and 'America'.

However, the play does fully carry through the potential of its satire of exploitative and fanciful aspects in film-making. Comic elements partly depend on McDonagh himself deploying a kind of exaggeration not dissimilar to the overheated melodramatic film script he satirises in Scene Seven. Names given to characters in themselves suggest overblown stereotypes, and in performance seemed 'larger than life', especially Aisling O'Sullivan's interpretation of the rather over-written Helen. Both Helen's scatological language and her violent suggestive behaviour could be read as indicative of the grotesque and carnivalesque body, but the extent to which her role is subversive is limited. It is also difficult to decide whether Johnnypateenmike's over-laboured story-telling strategies can possibly claim to be read with postmodern irony. Similarly, Bartley's obsession with American sweets, which he demands the aunts stock in their impoverished shop, could have been more effectively used to critique the 'Dream of the West.' That locals stare at cows for entertainment, and one aunt takes comfort in talking to stones, lightly touches upon the isolation of rural life, but again only up to a point.

There is a brief mention of history – when Helen throws eggs at Bartley as a game of England versus Ireland (ibid., p. 51) – but the link between this colonial relationship and the literal and cultural poverty of Inismaan is not made.

The grotesque body of Cripple Billy does, in himself, embody the drive to migrate and how it may be fed by fantasies of 'Otherness'. He at least is shown as aware of the split between his own identity and that identity as represented, whereas the others are constructed virtually as caricatures not too distant from the 'Stage Irishman' so disparaged by the Irish National Theatre's founders. Just as *The Man of Aran* does not convey an authentic picture of island life, neither in a different way does McDonagh's – but more to the point, neither does the 'Dream of the West' accurately depict America. In reply to Bartley's need to hear how great America is, because his emigre aunt lives there, Billy merely replies, 'It's just the same as Ireland really, full of fat women with beards' (ibid., p. 65).

The question of authenticity in images of Ireland is germane to the difference between *Stones in His Pockets* and *The Cripple of Inismaan*. The latter, despite its slight element of deconstruction, is more like *Man of Aran* and *The Quiet Man*, and thus to some extent a product of that link between nostalgia and aspects of the postmodern condition that Bauldrillard claims arises

> When the real is no longer what it used to be, nostalgia assumes its full meaning. There is a proliferation of myths of origin and signs of reality; of secondhand truth, objectivity and authenticity (Baudrillard, 1983, pp. 12–13).

Colin Graham's analysis of authenticity in the post-colonial Irish cultural context ((eds) Graham & Kirkland, 1999, pp, 7–28) sets up three categories: old authenticity, new authenticity and ironic authenticity. The ambivalent, borderline position of Yeats – and presumably other Literary Revivalists – is seen as an example of old authenticity, which both vindicates the colonized while granting special status to the collector and re-discoverer of 'authentic' texts (ibid., p. 18). Grahams's exposition of new authenticity links it with post-colonial market forces, which have both mythologized and celebrated 'the authentic' through a 'process in which the authentic is commodified, reproduced and retold' (ibid., p. 21) – as, for example, through tourism, 'heritage products' and certain kinds of TV programmes, especially those shown in Britain. This phenomenon, which elsewhere Ruth Barton has called the '*Ballykissangel*ization of Ireland' (2000, p. 413ff), Fintan O'Toole has linked with 'Disneyfication' (1994, p. 37). The Irish-American production of the two cited films can – more understandably in their diasporic context – be seen as examples of this 'new authenticity'. McDonagh's much more recent play seems a less excusable example in its double-tongued excess, its failure to provide an adequate context and in its mostly conservative dramatic form, which is closest to Kearney's category revivalist modernism.

Stones in His Pockets, on the other hand, does give a deconstructive image of Ireland in tune both with Graham's term 'ironic authenticity' and Kearney's radical modernism. Marie Jones draws upon immediate experience from within communities. The plot centres on a present-day visit to County Kerry by a film company, which pays extras forty pounds per day. These include Charlie Conlon from Ballycastle, whose small video shop went bust in competition with a large firm ExtraVison, and Jake Quinn, a local who migrated to America, but has returned, unsuccessful. The stage is bare except for a large black metal chest likely to contain filming equipment, and a row of tattered boots upstage, reminiscent of *Waiting for Godot*. Two actors, Conleth Hill (whose performance won the 2001 Olivier Best Actor Award) and Sean Campion, play these two main roles and all the other 11 characters. With minimal props and rare changes of token costume items they relied on brilliant use of minimal gestures, rapidly changed body language and positioning to show different characters and locations. The emphasis on their physicality embodies carnivalesque qualities cited by Bakhtin, especially as they represent as it were the carnival crowd, making the audience aware of their membership of

> a continually growing and renewed people [...] (through) festival folk laughter (which) presents an element of victory [...] over all that oppresses and restricts'.
>
> (ed.) Morris, 1994, p. 210)

Similarly, for all the Irish extras the break from daily mundane routine, the assumption of role, feasting from the film-makers' mobile catering wagon and the mixing of individuals from different classes and status, all contribute to the sense of carnival in Bakhtin's sense. These extras include Jake's mother's third cousin Mickey, 'the last surviving extra from *The Quiet Man*' (2000, p. 11), who puts his knowledge of the process to good effect, although he has never done more than drink away his pay for extra roles over all the years since. The inner film, romantic in 'heritage style', has an American star Caroline as heroine, and is directed by an Englishman with two assistants – Simon, 'an ambitious Dublin 4 type', and Aisling, a pretty girl with interest only in those above her. Events cover the shooting of various scenes, and the first half includes Caroline's seductive attempts to use Jake as a dialect coach, but he is uncertain about his fate as a 'sex object with an accent' (ibid., p. 33). When I saw the production, the audience gasped at Simon's coercive line to Jake – 'Look mate, Miss Giovannis has a habit of going ethnic. Helps her to get into the part' (ibid., p.30). The first half ends with news about Sean, a young unemployed local lad, given to drugs and despairing because his farmer father went bankrupt. Refused a role as extra, and kicked out of the pub at Caroline's request, he has drowned himself. Seen at a distance, he walked into the water once, came out to load his pockets with stones, then walked back.

The second act starts with Sean's friend Fin, an extra, talking about Sean's longheld ambition to migrate to America – a dream kindled by their experience of watching another film shot when they were children. Local people refuse to carry on filming on the day of Sean's funeral, despite the hostility of the film-makers – who eventually capitulate, allowing them a brief 'dry' time off, whilst Caroline cunningly arranges for flowers redundant from the set to be delivered to the church. Throughout, Charlie, who has written a rather dubious film script – with cliché aspects suggestive of 'new authenticity' – has unsuccessfully tried to get someone from the industry interested. He and Jake finally realise they do have a right to tell the story of Sean's suicide,

> so the stars become the extras and the people become the stars [...] in that way (it becomes the story of) all the people in the town' (p.54).
>
> (p. 59)

Resisting the English director's advice that their idea is not commercial, in their self-styled role as 'Canvas Productions', they finally start imagining their non-sentimental starting point – earthy shots of cows.

Throughout, partly evocative of Brecht's Epic style, the whole notion of representation is shown ironically, constantly revealing filmic artifice which creates a so-called 'authentic' image of 'Ireland', and deconstructing the inner film's 'new authenticity'. When Jake and Charlie are meant to be looking like dispossessed peasants as Rory the hero appears:

> It's not us they want. Its the Blasket Islands [...] they'll get a big shot of the Blaskets and the peasants, then Rory comes out of the hill behind us like he is walking out of the sea. When he has his line, the lot of us disappear, even the Blasket Islands.
>
> (2000, p. 15)

The outer play consistently puts socio-economic conditions and aspects of the people/extras' historical colonial context back through the theatrical image, fighting against the soft, romantic approach of the inner film. Jake points out,

> Mickey has watched his whole way of life fall apart around him [...] and now all it's worth is as a backdrop for an American movie [...] he depends on their forty quid a day and then he lives in hope of the next one.
>
> (ibid., p. 45)

Caroline is annoyed during her 'language-coaching' sessions with Jake, when he tries to explain that Maeve the heroine from the big house would have been English-educated and would thus not have spoken like the peasants. Charlie suggests her terrible Americanised accent does not matter because '[...] been that many film stars playing Irish leads everybody thinks that's the way we talk now' (ibid., p. 14). The issue of authenticity is often foregrounded, as when extras are put to turf-digging and similar activities like passive resistance to eviction, which they consider comically inappropriate, but the film-makers think are typical of such 'simple, uncomplicated, contented' people (ibid., p. 15). A sharp contrast with the inner film's cosy ending – shot at the start of Act Two, when the peasant hero gives back to the people the land he has gained through marrying Maeve – is gained by its ironic juxtaposition with the following discussion about possible reasons for Sean's suicide. Jake suggests 'Maybe he looked at me and realised there was no American Dream' (ibid., p. 38).

These indications of history's significant presence might suggest the play is sombre – on the contrary, it is extremely funny and energetically celebratory in moments such as when Irish dancing is performed almost as a challenge. Fluid form, which includes flashbacks in time as well as flexibility of locations, is particularly suggestive of the notion of theatrical space as a heterotopic site – since the play's radical modernist strategies work ironically to challenge the new authenticities of a certain kind of film representation of Ireland. The exceptional dexterity of the actors, in slipping from role to role, facilitates the carnivalesque body's suggestion of a multiplicity of potential identities, but it also has a particularly profound function here. In his exposition of heterotopia, Foucault uses an analogy of looking in the mirror, saying, 'I discover my absence from the place where I am, since I see myself over there [...]', then showing how this gaze moves back from the image in the virtual space beyond the glass, returning to himself, 'I begin again to direct my eyes towards myself and to re-constitute where I am' (1986, pp. 22–7). Thus these extras, who have begun to feel estranged from themselves by the fabricated, simple, single subjectivities of the film-maker's version of their identity, are now able to begin to reconstitute themselves by taking charge of their own story, and reversing, in Bakhtinian carnival style, the hierarchy of teller and told. Their comic energy triumphs, in a sense bringing back the dead to life ((ed.) Morris 1994, p. 210). Thus the transformative potential of the heterotopic space is, like Homi Bhabha's 'Third Space', an opportunity for the mobilisation of destabilised identities (1997, pp. 191–2, p. 206). An audience, observing ways in which multiple identities are created through performance, is also made more aware of those other ideological forces that impinge on both cultural identity and representation in the post-colonial context. Both *The Cripple of Inismaan* and *Stones in His Pockets* are comedies, yet choose drowning using stones as a sign of despair. The former falls into the trap of using melodrama in a way which evades history; the latter draws upon history in a drama that is a comic tour de force, but also challenging of those globalising forces that lie behind the film and tourist industries and are linked to diasporic dreams. Like the beer advertisement analysed by Graham, *Stones in His Pockets* in its 'ironic authenticity' deconstructs and subverts the claims that are made by 'the excesses of populist versions of restored Irish authenticity' (1999, p. 25).

The influence of film representations of America can be seen in Murphy's *The Blue Macushla* (1980), which draws upon American detective stories, while Barry's *White Woman Street* (1992) and Moxley's *Danti-Dan* (1995) both involve aspects of the American western. In *Whistle in the Dark* (1961), Murphy had explored the fate of Irish emigrants in Coventry. O'Toole states this play shows 'emigration is not a solution to Irish problems, but merely the sharpest indicator of how severe those problems are' (1997, pp. 171–2). Written almost twenty years later and premiered by the Abbey, *The Blue Macushla*, like *Danti-Dan*, reveals yet critiques aspects of the influence of American films upon issues of cultural identity. Murphy has claimed that in using the structure, characterisations and style of American gangster movies he realised this was 'an apt metaphor for a play about Ireland in the 1970s' (1992, p. xxi). He suggests the violence of gangster bombings, killing and corruption rackets is not too distant from the activities of the Garda Special Branch, the IRA, political conspiracy and corruption scandals, which he found on his return to Dublin after several years' absence. Murphy skilfully weaves gangster discourses with local idiom in a very complex plot that involves double-crossing and multiple identities. Eddie is forced by a protection racket to allow his eponymous club to become the cover for off-stage violence. Events follow a flashforward scene in which Eddie is seen to shoot a hooded figure – finally revealed not to be a priest but another agent (the Countess), substituted for the original victim by Danny, supposedly Eddie's financial partner. Strongly satirical comedy contrasts Eddie's butter and pig-smuggling to the North with the serious violence of politically-driven protection racketeers. It also subverts cliches of the genre through touches such an agent as Santa-clad piano player, a seemingly promiscuous kindly singer, and the use of names with historical connections, such as Countess and Mountjoy. Eddie's oath of allegiance to the organisation combines gangsterism with politics – 'I, number 19, division Dublin North-Central [...] will actively seek to establish and defend a United Ireland' (1992, p. 159). O'Toole's detailed analysis foregrounds the play's attack on militant Irish Nationalism but emphasises the significance of the Saint Patrick's Day setting, as this is an American creation 'celebrating an American conception of Ireland which has been accepted as home grown' (1994, p. 162). Eddie's bar serves green Guinness decorated with Irish and American flags, while the doorman praises American-style majorettes off stage. The singer's repertoire tends to the sentimentally Irish, redolent of Graham's new authenticity – including an emigration reference, 'Off to Philadelphia in the morning', rendered in a cloth cap and 'Paddy' costume. The club's pleasures and gangster strategies suggest the playboy approach to living rather than the puritan ethic. Eddie's own struggle up from urban slums, though based on corrupt and criminal activities, contrasts sharply with both the Irish song's dream of America – 'I'm sick and tired of working [...] I'll be of ta Cal'forny, an 'stead of diggin' praties I'll be diggin' lumps o' gold' (1992, p.185) – and with the Americanised romantic dream of Ireland. The play thus exposes the dangers of both these dreams, as well as the search for cultural identity during a time of upheaval and confusion.

Danti-Dan, written for Rough Magic, directed by Lynne Parker, toured Ireland and played at London's Hampstead Theatre in 1995. Set during a 1970s summer in a village ten miles from Cork, it explores adolescent sexuality in a rural context within a realist form, merging aspects of the 'Dream of the West' and the 'Dream of America'. Where *The Blue Macushla* drew fully upon gangster movie discourse, Dan, the fourteen -year-old innocent with a functioning age of eight, uses virtually nothing but the language of the Western genre. He rides Trigger, an imaginary horse, making the sound 'danti-dan' for the hoof beats, twirling a lasso and 'shooting'. Near an erratically working phone box, a river bridge and a monument for Nationalist dead, two early-teenage girls Cactus and Dolores idle their time away with Dan, sometimes joined by Ber, Dolores' older sister who works in the village stores, and her boyfriend Noel. Their uninhibited dialogue, more strongly suggestive of licence and sexual pleasure than even the Revivalist's dream of the West of Ireland, is strongly carnivalesque in Bakhtin's sense, making jokes about body parts, masturbation, intercourse, pregnancy, food, drinking and smoking and so on, and played out in a lively physical style. Noel, something of a playboy, is out of work, and reluctant to find it, despite pressure from Dolores. Paradoxically, the mentally disadvantaged and innocent Dan is meticulous in what he sees as his work, taking down car numbers in carefully ruled books, with something of the cowboy work ethic. Dan is delighted when Noel, teasing, suggests he will take Dan to America to 'ride the fucking range' (1999, p. 342), but that Dan will have to learn to play cards. Cactus, frustrated both by reading romances and kissing Dolores, later bamboozles Dan into what she calls 'playing poker' with her – but by her own invented rules, which involve sexual games and intercourse. She pretends the council will pay Dan for his car-recording books if he keeps their secret, so in his determination to get money to travel to America he continues her sex games until Dolores also tries them, and they are caught by the now-pregnant Ber. Cactus pushes the struggling Dan, desperate to retrieve his precious book, off the loose bridge coping. He falls to his (unseen) death. The claustrophobia of rural life for young people is well captured: apart from Cactus moving to Limerick, life continues normally, the Council is blamed for the faulty bridge. However, both plays, set in the 1970s, suggest there is something dangerous and perhaps immature in illusions conjured by the American dream, consonant with O'Toole's analysis that mythic America is 'enormously destructive for the Irish young' (1994, p. 77).

Premiered at London's Bush Theatre, Barry's *White Woman Street* (1992), although set in the American West in Ohio during Easter 1916, has a more progressive approach to identity in the context of migration. As elsewhere, Barry's characters seem on the far edge of history. The distant momentous Rising in Ireland impinges only briefly on events, as near the play's end, Clarke, a native American store keeper, remarks:

> Place there burning like Richmond, I hear. Some big mail depot or someplace. Fire and ruin in Dublin. Fellas put in jail and likely to be shot. Fighting the English.
>
> (1995, p. 175)

Trooper O'Hara, the Irish Outlaw, is shown to be not listening to this – 'That Right?'. The outlaws led by O'Hara are in different ways marginal characters with a sense of displacement. O'Hara himself is ironically aware that,

> Indian towns [...] put me in mind of certain Sligo hills and certain men in certain Sligo hills. The English had done for us I was thinking, and now we're doing for the Indians.
>
> (ibid., p. 158)

This acknowledgement of colonial dispossession is not the major point in this partly mythic Western tale, which centres on O'Hara's need after twenty years – before returning to Ireland – to revisit a small town where once the only white whore for five hundred miles around had plied her trade. The outlaws intend to rob a gold train, reputed to be loaded with local soldiers' Easter pay. Different reveries illuminate the history and thoughts of each outlaw, so the audience can appreciate how each one has been exiled from his place of origin. The troop includes a Russian with a Chinese mother, a black man, an American Indian, an Englishman from Grimsby, and Mo who was born into an Amish sect. A sense of drifting within the narrative seems analogous to the outlaw's wandering lifestyle, while the fading romance of the West is suggested by tensions between tragic and comic elements – such as their quasi-domestic routine. References to the harsh life that had prompted migration from Ireland are repeated in the harsh demands of life in America:

> those farms they left when food got scarce [...] pushing spades and hauling up axes in America digging and cutting like pigs and dogs, men whose children had froze in ditches [...] filthy flotsam of Ireland, letterless, stumbling, crooked men who had nothing to believe on but that there was one white woman here [...]
>
> (1995, p. 164)

Eventually it is revealed that the white whore is but a legend, and that, against audience expectations she had not been killed by O'Hara. Rather, on visiting this mythic supposedly-white woman, he found he had taken the virginity of a young Indian girl who immediately slit her own throat. Thus O'Hara has long borne a guilt which he sees as a murder – himself a colonised individual, he has operated as a coloniser of Others. Mo persuades O'Hara to lay his guilt to rest – 'What happen in America is like a rover flood, everything lifted and dragged away from its place' (ibid., p. 178). After their raid on the gold train fails, O'Hara dies as, encouraged by Mo, he imagines he sees again the hawthorns of home in Ireland. Shifts of time and location, and reveries were signalled by lighting changes, poetic passages and Shaun Davies' music. Kendra Ullyart's hyper-realistic set and touches such as the use of imaginary horses, took the play away from a strictly realist genre. Here the American dream is shown, like the myth of the woman, to be flawed, illusory and another kind of colonising activity. Nevertheless, the play looks to the future in presenting the brotherhood of uprooted and marginalised outlaws as a kind of paradigm of plural cultural identities.

If dreams that fuelled emigration are represented as misleading, so the reverse journey home, whether permanent or temporary, is often shown as disillusioning. For example, Pom Boyd's *Down on to Blue* (1994) for Rough Magic reveals how Joey, who has worked very successfully in America, finds on a visit home that her dysfunctional, downwardly mobile middle-class family makes it impossible for her to accept a good job back in Ireland. Katie O'Reilly's *Belonging* (2000), premiered at Birmingham Repertory theatre, shows how Maura (played by Eileen Pollock), living in Birmingham for twenty years, had avoided identification with Ireland – even getting her children elocution lessons to obliterate their accent – then decides to visit home. In adulthood, her son is indifferent to his cultural identity seeing himself as European, but her daughter is a professional Irish woman, an oral historian working on emigrant memories. Ultimately, 'Maura's face turns to scarred granite when she realises there is no home to go back to' (Lyn Gardner, *The Guardian*, 9 December 2000).

Similar disenchantment is found in Murphy's earlier play, *Conversations on a Homecoming* (1986), directed for Druid by Garry Hynes. Set in the early 1970s, it deconstructs what O'Toole's Introduction calls the 'long hangover from the 1960s' when the Republic began both to regret its American-rooted optimism and to face its own grimmer social realities, while the Troubles were increasing in Northern Ireland. Michael has returned from America to the White House bar in an East Galway town, where in his younger days he consorted with his peers, encouraged especially by energetic and idealistic JJ – who had seen himself as like John F. Kennedy, with the bar aspiring to be a kind of Arts Centre version of so-called Camelot. Michael waits in vain for JJ to return from a drinking binge. His old friends Tom, Liam and Peggy are in their different ways also a disappointment, while Junior, a successful business man, affects a slight American accent. This play undercuts both the dream of the West of Ireland and the American dream through its exposure of the frustrations of those left behind with what they feel is Michael's New York naivety:

> [...] American politics or business methods. Jesus, images, fuckin' neon shadows!
>
> (1997, p. 54)

Tom likewise castigates the sentimentality of what he calls 'the country and western system', trying to hold out for the possibility of change through creative cultural minority groups. After a drunken evening, Michael realises their past is gone forever.

Declan Hughes, Writer-in-Residence with Rough Magic from 1992, also acknowledges that as he grew up in Dublin in the 1960s and 1970s, Irish cultural influences seemed related to the past, but for TV, films and music 'to discover the present you looked to America' (1998, p. ix). Thus,

> [...] if you felt your cultural identity dwindling into a nebulous blur, well you believed that what you had in common was more important than what set you apart, you knew there were millions like you all over the world, similarly anxious to be relieved of the burdens of nationality and of history.
>
> (ibid.)

Hughes' *I Can't Get Started* (in (ed.) Burke, 1999) is a film-noir influenced exploration of the relationship between American writers Lillian Hellman and Dashiel Hammett. His *Digging for Fire* (1991) and *Halloween Night* (1997) – both in *Plays One*, 1997 – each dramatise the relationships among groups of friends, including a returned migrant. The identities of these young, mostly professional people seem almost typical of any First World town. David Grant of the Lyric, Belfast, suggested that paradoxically this 'less-Irish' quality may hinder Hughes' chance of productions outside Ireland (Interview). Hughes opposes 'museum culture' evident in reviving rather than re-writing classical plays (1997, p. xi). In *Digging for Fire* the destruction of an already-shaky marriage between Brendan and Clare is triggered by the return from New York of Danny, with whom she had an adulterous affair. At the group's drunken reunion, arranged by unaware Brendan, Danny's claims to be a successful writer are revealed as pretence, and the group's friendship as an illusion. Steve, a worker in advertising, foregrounds the problem of authenticity;

> You can have whatever you want if you can pay for it. Ireland as folksy little village or as fifty-first state in terminal stages of urban breakdown – all it is is goods and services.
>
> (1997, p. 39)

As a partly open ending, Danny decides to stay in Ireland – to 'dig for fire' in his writing – and Clare dances alone to celebrate her new autonomy.

Halloween Night is not as dependent on realist form as the earlier play. On tour at London's Donmar Warehouse, spooky festive decor, mainly red and black costuming, and a huge backdrop of Gericault's 'Raft of the Medusa' provided a mysterious atmosphere for a reunion set up by George in his seashore house. The professional/media group of friends – including previous lovers, gay and straight couples – await his return from America, becoming increasingly uneasy due to strange music. As they dance in masks, Todd, a Californian arrives to say that George, who was HIV positive, is now dead, then leaves. Unearthly music and lights at point impinge upon the wake-like party, which has – perhaps as George planned – become a catalyst, releasing dangerous undercurrents in the group's understanding of each other's identity. Further phone messages suggest both Todd and George are still in-flight and alive. As apocalyptic music accompanies the rising sea outside, threatening the unnerved friends, a further phone call states George did die in America. Huddled in a pose that mirrors Gericault's survivors, the friends wait as the storm door creaks open to let in light, a potential epiphany.

Two further plays that demonstrate the effect of migration on identity, and the impossibility of returning to the past and home as it was, both focus on problematic wakes. Pigsback's production of Joseph O'Connor's *Red Roses and Petrol* (1995) opened at Project Arts Centre, Dublin, then London's Tricycle. Murphy's *The Wake* (1998) was first directed by Patrick Mason at the Abbey. The first play explores the death of a father, the second of a grandmother: both show the disruptive effect of returning family members, whether from America or England. O'Connor's play explores the after-effects of the death of an academic librarian and poet, Enda. While one daughter Medbh has stayed in Ireland, the other, Catherine, flies in from New York with her lover Tom, originally from Galway, whom she met in an American club, the Rhinestone Cowboys. Son Johnny arrives later from London. At the start, while Moya and Medbh sort out books, the RTE closedown music accompanied by rural scenery is visible on TV, as if to emphasise the importance of image. It transpires that Moya has done her best to preserve the appearance of a happy marriage, despite a scandalous university rumour linking her husband with a much younger women. As she clings to happy memories – her first meeting with her husband in London and his regular weekend purchase of red roses and petrol – the siblings quarrel, their different perspectives offering competing versions of the past. Two short happy flashbacks to his courtship of Moya, and Enda's appearance as revenant on some videos of himself which he has left among his effects for his family, add further to this slippery sense of the past. In them he stresses his great pleasure in the rural West when hearing Irish spoken, and written in Douglas Hyde's collection of *The Love Songs of Connaught*. This volume Moya later consigns to the library, as one no-one reads (1995, p. 96). No one other than the family turns up at the wake, because Catherine discouraged them secretly, lest the 'other woman' appear. Eventually Johnny's knowledge of Enda's will reveals this young woman was actually Enda's daughter by someone else, after the birth of his other three children. Moya wants to keep this secret from her daughters so happy images of the past endure, and the play ends with her listening to Enda read a poem for her on video, while what may be herself and Enda (Catherine and Tom) are seen embracing. Both Johnny and Catherine had emigrated partly to escape what had been in reality a troubled family home, and it is clear they will never return to live in Ireland. Medbh cannot decide whether to leave for Australia with a friend, leaving her mother alone. Thus the play suggests motives for migration may be emotional as well as economic, enabling individual identities to escape suffocating yet idealised family life.

Murphy's *The Wake* also suggests sexual restrictions and family problems may lie behind such a bid for freedom. Vera O'Toole, a call girl in New York, comes back to a rural community to pay her respects to the dead grandmother who looked after her for eleven years of her childhood – she also wishes to be thought worthy by remaining family members. This dream is shattered not only by the fact that her grandmother lay dead for some weeks unnoticed, having fallen into the fire, but because there had been no wake, only an inquest. Vera's brother, Tom, had forbidden neighbourly visits to the grandmother, presumably lest claims are made on the estate. Remaining family members, really only interested in their businesses and material possessions, are worried about Vera because she had also inherited the Imperial Hotel, the 'jewel in the crown of family fortunes' (1998, p. 33) from her mother's will. Vera, something of a carnivalesque figure despite her inclination to depression and pill taking, approaches the problem eliptically. First she enrols the help of Finbar, a disreputable old lover – a bachelor brutalised and sexually abused by Christian Brothers as a child. Despite now having 'problems on both side of the Atlantic' (ibid., p. 42) because she has to forgo dubious sexual engagements in America, she subverts her family's plans to get the (closed) hotel via a rigged auction. She holes up there for a drunken few days with Finbar and Henry, one of her brothers-in-law – a drop-out alcoholic lawyer from a decayed Protestant Ascendancy background. The police, a doctor and Father Billy, are reluctantly drawn into the situation. Henry's enraged wife had suggested Vera should be locked up, as sometimes happened to brother Tom's mentally disturbed wife, Catriona. Eventually Vera meets formally with the family in the hotel, to 'finalise, bury everything and mark the occasion – with a wake' (1999, p. 88). This marks the death of her dreams of returning to what Henry has called this 'most distressing country' (ibid., p. 64). Vera explains how in her years away 'I used to think I was real because I came from here' (p. 87), whereas it is clear her family has rejected her for their own ends. After this evening of drinking and song, which partially seems to restore the disrupted family marriages, she reveals that she will relinquish her claim on the hotel – but in the last scene weeps bitterly, alone, over her grandmother's grave and the prospect of starting again, elsewhere.

A feeling of estrangement from home and identity may be found even with those who, returning, decide to stay. Set in the 1890s, Barry's *The Only True History of Lizzie Finn* (1995) explores this ambiguity around belonging. Barry's ancestor Lizzie, an entertainer, was brought up in Corcaguiney by her poor hardworking mother and travelling singer father. On tour with her friend Jelly Jane in Weston-Super-Mare, England, she meets then marries Robert Gibson, surviving son of a fading Protestant landlord dynasty, and moves to the big house at Inch in Ireland. Despite Robert's mother's initial shock, she learns to love Lizzie. Robert – something of a misfit, since he believes in equality 'under the clothes that history lends us' (1995, p. 35) – challenges the colonial attitudes of Lord Castlemaine and local society by revealing the truth about how his two brothers died fighting against the Boers, 'as nobly as people can die when your general is a fool and the cause is unjust'. Further, the third brother 'died of drink in Cape Town. That's how he took the news' (ibid., pp. 51–2). Once society learns Robert, disillusioned, changed sides to fight for the Boers, the rector prevents his mother from attending chapel and she drowns herself. Hypocrisy at her funeral finally drives Robert and Lizzie to leave Inch for Cork. There they can be, as it were, foreigners, living unconventionally, evading the landlord class's restrictive expectations and colonial values colonial values. Clothing is a particular signifier of character and position: Lizzie's energetic bohemianism is indicated by gold stars placed under the crotch of her stage cancan knickers; Ro9bert shocks the gardener by wearing rough working clothes to repair fences. In tune with Barry's view of human history, suggested doubling across class roles suggests humanity is more significant than social distinction:

> There'll be nobody in the world to remember us, child, and all that will remain with us is an echo, a strain of dancing music, and the memory of a man that loved his brothers and his people ... So what odds where we are?
>
> (p. 64)

This notion that geography is not intrinsic to identity looks forward to the potential flexibility and multiplicity of identities (discussed further Chapter Seven) as especially significant when more people are returning to Ireland, as a whole, in greater numbers than ever before.

As this chapter's opening remark by Gibbons suggests, the diaspora's long-term existence has influenced the need to create images of home that are appealing rather than uncertain and destabilising. Hence, drawing more strongly upon indigenous dramatic traditions of comedy and melodrama is evident in television programmes set in Ireland and recently shown on British television. Consequent difficulties of using these genres can be seen through brief reference to three comedy programmes: *Ballykissangel* (series from 1996–2001) commissioned by BBC Northern Ireland and World Productions, financed due to John Birt's regional quota system; *Father Ted* (three series from 1995–1998) made by Hat Trick Productions for Channel 4; and *The Fitz* (one series, 2000) created by Tiger Aspect and made in Northern Ireland. Typical of the English idea of Ireland as 'a locus of play, pleasure and fantasy' (Eagleton 1995, p. 9), all three programmes have a rural context. *Ballykissangel*, as the credits reveal, has an idealised local setting, which has brought a location tourism boom to the County Wicklow village of Avoca, where it is filmed. Craggy Island, the site of Father Ted's presbytery, uses a postmodern concept of space and time, as the island shrinks or stretches according to the demands of a particular storyline, and any sense of the real is in tension with the series' surreal and absurd aspects. *The Fitz*, apparently set in a rural home, uses deconstructive techniques and media-linked framing devices linked with the absurd – the border between the Republic and the North runs through the house to humourous effect.

Drawing upon Robin Nelson's terms (1997), it is possible to consider *Ballykissangel* as an example of 'formulaic realism', blurred with soapy melodrama; and *Father Ted* as an example of 'critical realism', which is 'the agent of the grotesque' – it is satirical, deconstructive and carnivalesque in its absurdity. (1997) A new term, 'disruptive realism', could cover elements of the grotesque, excess and uncanny repetitive structures which feature in *The Fitz*. Representation of gender identity also differs. In *Ballykissangel*, many males seem emasculated and, despite some hints of the 'new woman', characters tend towards stereotype, fulfilling typical rural roles as indicated in Kavanagh's novel *Tarry Flynn* (1948). *Father Ted*'s priests are in different ways also emasculated males; females, such as the rock star or the 'Lovely Girls' contestants, provide exaggerated images (as in *Rock a Hula Ted*), while Mrs Doyle the housekeeper seems drawn with a misogynist touch. The Fitz family embody a wild range of stereotypes, but with elements of fluidity and camp, especially cross-dressing 'Ginger Jean, the Fenian Queen'. *Ballykissangel*'s slight criticism of the church and minor economic reference is far outweighed by *Father Ted*'s consistently anti-clerical satire, while *The Fitz* does contain some critical one-liners about both church and political events. *The Fitz'* border setting includes religious satire and the constant prowling of inefficient patrolling soldiers – with references to the liminal effect on issues of political and gender identity, which in one episode are linked to the cultural position of a deserting black soldier. *The Fitz*, with every family member named after the American presidential Kennedy family, does perhaps provide more potential for flexibility of identities – including the ceaseless job opportunities for daughter Teddy. From a postcolonial perspective, it might be possible to claim that both *Father Ted* and *The Fitz* draw upon what Mercier (1962, pp. 48–9) claims as the dionetic quality of traditional forms of Irish humour, the macabre and the grotesque, associating the former with terror and the fear of death, and the latter with a dread of the mysteries of reproduction. In this sense both programmes draw more strongly on indigenous culture from within, while *Ballykissangel*'s new authenticity reflects merely the surface of Irish culture represented as a commodity for the outside viewer, tinged with bland elements of traditional melodrama. Traces of ironic authenticity may be present in *Father Ted*'s use of metatheatricality to expose cultural flaws. *The Fitz* also uses TV itself intertextually and deconstructively, while both re-work cultural reference – for example Synge's *Shadow of the Glen* (1904). In different ways all three dramas persist up to a point in constructing the Irish as Other, particularly as television discourse, when drawing upon types of realism within comedy genres, tends to keep history offstage, rendering 'The forces which shape these men and women [...] invisible and opaque' (Eagleton, 1995, p. 313). All three programmes problematically flirt with the danger that their use of stereotypes may be misread. Potential radical effects of absurdist excess in *Father Ted* and *The Fitz* may be limited since postmodern ironic readings cannot be guaranteed. Nelson underlines Caughie's reservations about such TV programmes, which although they may open

> identity to diversity, and escape the notion of identity as a fixed volume [...] (do not) do it in that utopia of guaranteed resistance which assumes the progressiveness of naturally oppositional readers who will get it in the end.
>
> (1995, p. 55)

Hence, despite the first two programmes' popularity and the more cautious reception of the third, it is difficult to gauge how such images of Ireland are received by the general British audience in comparison with audiences with Irish cultural roots and experiences.

No plays discussed here treat migration – whether to England, America or elsewhere – as unproblematic. Chapter Seven, in analysing the perpetuation of stereotype characters and communities, evaluates further the issue of authenticity in drama in relation to dramatic form. Certainly nostalgic images blended with simplistic notions of identity, particularly those linked with the rural landscape in the post-*Quiet Man* mould, are no longer representative of the new more urban Ireland, which according to Fintan O'Toole, is

> essentially the same kind as the place in which the exile now lives, all the more so because memory itself is now saturated with globalised media images.
>
> (1997, p. 174)

How crucial therefore, is the exportation of images that have a strongly deconstructive ironic authenticity and which celebrate the hybrid multiplicity of postcolonial Irish identities – of a culture that, due to the diaspora, knows no borders.

7 Conclusion: From Hearth to Heterotopia

Throughout this book, a central concern has been the ways in which contemporary Irish drama both reflects and constructs cultural identity within a post-colonial context. From the inception of the Irish National Theatre, the complex role of realism, evident in the poetic approaches of Synge, the formal and stylistic experiments of Yeats and the expressionism of O'Casey, is especially significant. Subversion of realism's formal and ideological limitations deploys strategies that range from use of the mythical and magical to disruption of linear time, from the splitting of subjectivity to the deployment of the carnivalesque and grotesque. Further, liminal, potentially heterotopic space is evoked, especially through the deployment of the actor's body in performance. In this sense, even the 'hearthside' of apparently fourth-wall settings across the period, whether the cottage of Yeats' *Cathleen Ni Houlihan* (1902) or the pub of McPherson's *The Weir* (1997), perform overtly and/or subtextually the transformative potential of identity. In the first case, the transformed identity of the old lady, who yet is young with the walk of a queen, is clearly Nationalist within a colonial context; whereas the latter evokes more personal issues in tune with the evolution of more fluid, hybrid identities, while acknowledging the specificity of the rural Irish economy. Across themes of memory, history, politics and gender, the more recent plays and productions discussed in previous chapters have varied in the extent to which they have disrupted realism or challenged fixed notions of cultural identity. Similarly, the relationship of these plays to cultural traditions is variable – making the overlap between postcolonial and postmodern elements sometimes difficult to differentiate. Further, audience reception remains ideologically complex, according to communities of interest or location (see Bennett, 1990).

Authenticity remains a central issue, especially because, as the closing of Chapter Six indicates, saturation by homogenising globalised media images is a threat even to the emergence of a more flexible and diverse concept of cultural identity. Not only is America 'viewed both as a producer and consumer of Irishness' (Graham, in (eds) Graham & Kirkland, 1999, p. 24) but as the major agent of global capitalism it floods Europe and the rest of the world with its companies, products and values. For Graham, the 'old authenticities [...] re-established after colonialism [...] rife for deconstruction by a globalized context' (ibid., p. 27) foreground problems associated with more recent attempts to represent authentically the newer and more radical urban, technological context of contemporary Ireland, since such representations have already begun to destabilise outdated traditional rural images. It is to be expected then that some contemporary Irish drama would show an ironic awareness of both old and new authenticities in terms of context and identity. Discussion of plays and performance here is therefore informed both by Stuart Hall's consideration of the importance of not merely 'after' but 'going beyond' the colonial situation (in (eds) Chambers & Curtin, 1996, p. 253), and by Kearney's views on postnationalist Ireland (1997). As the latter suggests (pp.

100–1), the present internationalisation of Irish art, literature, music, dance and drama has a very long history rooted in successive waves of migrating Irish men of letters to the European continent, even as far back as the seventh century. In contemporary times the relationship with the European Economic Community has further strengthened two-way traffic, which, according to Kearney (ibid. p. 100), James Joyce once expressed as a determination to 'Hibernicise Europe and Europeanise Ireland!'.

Acknowledgement of intertextual and intercultural elements in plays by Friel, Murphy and McGuinness follows Christina Reid's critiques of old stereotypes. Detailed analysis of plays by Jones, Bolger and Lynch then indicates more flexible approaches to identity, and an engagement with European concerns. Confrontation with political change in Mitchell's *The Force of Change* (2000) and a forward-looking play set in the past, McGuinness' *Dolly West's Kitchen* (2000), provide a formal contrast that epitomises the hearth-to-heterotopia dynamic. Finally this debate is extended through reference to the increasing significance of carnivalesque and physical theatre, performed by a range of groups such as Macnas, Barabbas and Tinderbox. Community contexts such as Dock Ward Company in Belfast and Mercier's Passion Machine in Dublin are included – as are issues around the publication of new writing. All this evidence illustrates the movement of contemporary Irish drama beyond the context of the colonial past and into a wider future.

Christina Reid's *Did You Hear the One about the Irishman* (1985) and *Clowns* (1996) both include aspects of stand-up comedy to deploy ironically what Graham terms 'the double-edged use of a stereotype' to challenge prejudices rooted in both old and newer authenticities (in (eds) Graham & Kirkland, 1999, p. 25). Originally produced in America during a Royal Shakespeare Company tour, revised at London's Kings Head in 1987, the former is subtitled 'a Love Story'. Set in 1987, it follows a doomed cross-sectarian romance between Allison, a middle-class Protestant, and Brian a working-class Catholic, which moves in Epic-theatre style between two households, also showing both families visiting friends and relatives – political prisoners in Long Kesh internment camp. As Brian says;

> So, where else in Northern Ireland can a Provie wife and a UDA wife take a long look at each other and realise they're both on board the same sinking ship. Common ground, common enemy.
>
> (1997, p. 77)

The whole is metatheatrically framed by outfront speeches from an Irishman and a Comedian, which provide a deconstructive Brechtian commentary. First, the Irishman reads out to ridiculous effect a list of items family visitors are allowed to take to the prison. The Comedian's typical jokes include,

> We are now approaching the city of Belfast. Will all passengers fasten their seatbelts and turn their watches back three hundred years.
>
> (ibid., p. 69)

His stereotypical jokes about Irish stupidity become increasingly racist during moments when he appears spotlit, usually in parallel with the Irishman. Jokes about religion and political violence on both sides of the Northern Ireland conflict – are intermingled with others like 'Grow your own dope. Plant an Irishman today' (p. 88). Problems, such as poor employment prospects for Catholics, include Allison's father's failure to breach this prejudice by employing a Catholic worker – as well as their Protestant family's silence about a Catholic great grandmother. Following a tender, joking scene between the lovers who plan to wed, the Irishman reads out and then tears up a news bulletin, which reports their death, but also includes references to Allison's uncle, a Unionist politician, and Brian's brother who is serving a life sentence for terrorist offences. Making some vicious jokes about death and the Irish – the Comedian provokes the Irishman into asking, 'What do you call an Irishman with an machine gun?'. To which the weary answer is 'Sir'. Juxtaposition of these blinkered, stereotypical jokes, in contrast with warm human relationships and socio-economic details subtly woven into the narrative provides an effective and balanced critique of the wider situation, which could prompt audience members to consider their own potential complicity in the Troubles.

Clowns, first performed in Richmond's Orange Tree studio, London in 1996, is set in Belfast on the eve of the IRA Ceasefire in August 1994. Four main characters, Sandra, Maureen, Arthur and Tommy were all in *Joyriders* (1986) discussed in Chapter Three. The old Lagan Linen Mill, then a site for Youth Training, is now a modernized open-plan shopping mall. Although a kind of peace seems imminent and the economy is improving, ambivalence about issues of cultural identity continues. Much new investment is British – and the characters discuss their vulnerability to bombing in contrast with such attacks on the British mainland. Because of the IRA ceasefire, Sandra has returned to Belfast for the first time since her friend Maureen was shot while protecting her joyriding brother. While in London, Sandra has become a stand-up comic, using the persona of her dead friend Maureen. Maureen functions both as a revenant seen only by Sandra, and somewhat as Sandra's 'split self', erupting into scenes. She appears not as ghostly but rather a romanticised 'village girl', who also uses a traditional blessing (1997, p. 289). A rainbow flower arrangement around idealised statues of a mill worker and child in front of the now-successful Arthur's restaurant are not, as it seems, a memorial to her. Such contrasts between illusory images of Ireland and reality are further carried through in the tension between Maureen's singing about the Mountains of Mourne and the actuality that her father, grandfather and great grandfather, who sang this when on the dole in Belfast or as casual labourers in London, never saw the scene. As the stand-up voice and persona, Maureen/Sandra's double-act jokes have a double edge – but unlike the earlier play, where only in the last moment was there a reversal of power, here the colonised take back their own power, most notably on meeting English ignorance:

> I mean, why should Bernard Manning and Jim Davidson make a living out of slagging off the Irish, when we can do it better ourselves. And with more wit and style. That's the real joke. You forbade us to speak our own language. You forced us to speak yours and we took it and turned it into poetry ... There were these thick Paddies [...] Joyce, Beckett, Behan, Synge, Yeats and O'Casey ... and there were the great *English* dramatists Wilde, Swift, Shaw, Sheridan, Congreve, Goldsmith ...
>
> (ibid., p. 306)

Future possibilities are variously signified: Arthur's mother Molly is a mature student, Tommy the communist has become a loosely socialist 'crusty' – but Johnny, Maureen's brother is now a drug dealer – offering the 'ultimate joyride', 'the only scene in Belfast that has nothing to do with religion, class or creed' (ibid., p. 330). Stand-up routines, including echoes from the earlier play, become increasingly savage, attacking the religion and leaders of both sides, ridiculing extremism and bigotry by overt statement and implication. They also express Sandra's ambivalence about her home city. As Act One closes, Maureen takes her hands from her abdomen and releases blood stains – jokingly she states that uncertainty about whether the gunman was Catholic or Protestant caused her to claim Jewish identity. Nevertheless he shot her, having just heard that two thousand years ago the Jews crucified Jesus. Raising a glass to the audience Maureen says, 'People have died for less' (p. 320). During the second Act Maureen, after quarrelling with Sandra, delivers particularly vicious invectives culminating in 'Either way you die blamin' it on the other side' (p. 326). Finally Sandra relinquishes her traumatised past, losing her sense of doubleness by accepting both Maureen's death, and the present potentially positive changes in Northern Ireland. The play closes in December 1994, where her London routine now includes jokes about the second, that is, Loyalist, ceasefire. She celebrates Irish resilience as a nation of comedians offstage – as in an overheard bus-stop conversation:

> You wait twenty-five friggin' years for a cease-fire and then two come along one after the other.
>
> (p. 343)

These plays both use the performance time/space of the stand-up comedian as a borderline/heterotopic site where different perceptions of cultural identity are disputed, and old monolithic sterotypes challenged.

Both Reid's plays can prove unsettling for an audience because the jokes undermine all positions, but the second ends with the possibility of hope. Since the play was written such optimism has been further endorsed by the Good Friday Agreement published in Belfast on April 10th, 1998, approved by both British and Irish governments and most local political parties, and reviewed under the aegis of former US Senator George Mitchell (September to 19 November 1999). Although consequent negotiations have been marred through fatalities caused by maverick groups on both sides and other problems, the Peace Process continues somewhat unsteadily. Coulter considers the Agreement to be reactionary in the way it seeks to validate equally the conflicting ideologies. He suggests that,

> Rather than choosing to venerate or simply accept unionism and nationalism we should be attempting to deconstruct and transcend them. Perhaps the time has come for a little more parity of *dis*esteem.
>
> (1999, p. 257)

Clowns especially performs this disesteem, while the notion of moving beyond these competing ideologies and identities accords with the greater openness towards Europe shown in much recent Irish drama on both sides of the border. The Republic's membership of the European Economic Community since 1973 has been further enhanced by its participation in the Economic Monetary Union since January 1999, though an Irish referendum in 2001 voted against further EEC Enlargement. Kearney also goes beyond the confines of past cultural identities by contextualising his vision of postnationalist Ireland within a 'federal 'Europe of the regions'' (1997, p. 77), suggesting such a European framework might eventually provide space for a solution of the Northern Ireland situation, which he sees as not dissimilar from 'other social and community conflicts mounting in intensity and range throughout the Greater Europe of the 1990's' (ibid., p. 76). Throughout, Kearney stresses 'every nation is a hybrid construct and imagined community which can be reimagined again in alternative versions.' (p. 188) This concept is, as previously discussed, very much in tune with Bhabha's notions of the significance of hybridity and the 'Third Space'.

Intertextuality is widely considered to be a feature of postmodernism, but in terms of intercultural relationships it is important to consider the dynamic of power relations between any two or more cultures, where there is an exchange of ideas. Pavis defines intercultural theatre as creating 'hybrid forms drawing upon a more or less conscious and voluntary mixing of traditions traceable to distinct cultural areas', but his hourglass model of intercultural transfer does not satisfactorily embrace the nature of colonial or postcolonial conditions (1992, pp. 4–7). Intercultural relationships between Ireland and Europe can be seen as a more egalitarian two-way traffic than the past imperial relationship between Britain and Ireland – which in cultural terms is nicely indicated by the quotation about writers and language from *Clowns* (p. 306), cited earlier in this chapter. In carrying further the European influences cited by Worth (1978) and indicated in this book's Introduction and throughout, contemporary Irish drama can be seen as 'going beyond' the old confines of postcolonial influences and reaching out towards the future, somewhat in tune with Kearney's postnationalism. The Irish specificity of ways in which such elements are re-worked avoids any claims for 'universality', a term which blurs both cultural difference and obscures the ideological implications of power relationships.

Beckett, who spent most of his life in France, writing his first plays in French, is a prime example of a playwright whose work reveals strong European influences whilst maintaining signifiers of Irish culture. His 'Theatre of the Absurd' has affected other Irish dramatists, including some discussed here (See Roche, 1994). European philosophical ideas, such as Heidegger's concern with the brevity of human life, Sartrean existentialism, Cartesian dualism, and post-Freudian notions of the self can be traced in the modernist concern with the nature of identity that occurs in Beckett's plays such as *Krapp's Last Tape* (1958), *Breath* (1966) or *Not I* (1972). These ideas follow through to the concern with the split self, doubling strategies, and fragmentation of time typical of postmodernist dramas previously described. As suggested throughout, Friel, Murphy and McGuinness also use dramatic strategies that are suggestive of European theatre practitioners such as Brecht, Artaud and Antoine or dramatists such as Ibsen, Chekhov and Pirandello. Thematically their plays also contain many intertextual references to European culture in terms of art, music and philosophy. This book, with its emphasis on new writers, has only discussed a relatively small selection from these three canonical writers, but a few indicative examples follow.

Friel has written a version of Chekhov's *Three Sisters* (1981) and an adaptation of Turgenev's novel *Fathers and Sons* (1987), while McGuinness has adapted Ibsen's *Peer Gynt* (1988) and *A Doll's House* (1997) as well as Chekhov's *Three Sisters* (1990) and *Uncle Vanya* (1995). Both writers have used classical allusion – for instance, McGuinness in *Carthaginians* (1988) explores the events of Bloody Sunday, 30th January 1972, in Derry through reference to the Roman Empire. In *Mutabilitie* (1997) he mingles Celtic myth with a staged vision of the Fall of Troy as well as Elizabethan history. Friel's *Living Quarters* (1992) subtitled 'after Hippolytus' combines a re-staging of the Phaedra myth with metatheatrical strategies, which echo Pirandello's *Six Characters in Search of an Author* (1921), including a somewhat Brechtian-style narrator. Other plays by McGuinness – his *Innocence* (1986) and *Mary and Lizzie* (1989), incorporate characters from European history: the artist Caravaggio in the former, and Marx and Engels in the latter. Murphy's plays also draw upon European elements, as for example the European fairytale elements within *The Morning After Optimism* (1971) and the Faust myth in *The Gigli Concert* (1983), in which the Irishman enlists the help of JPW, a mysterious English upper-middle-class Mephistophelean figure in his attempt to emulate the Italian singer. O'Toole's analysis of this play suggests the growing relationship between these two men is suggestive of two halves of a Jungian integrated personality (1994, p. 225). A more complex, detailed analysis of most of these plays, which assesses them differentially in terms of a process of socio-cultural decolonisation, which through their European dimensions transcends the British influence, can be found in Nikolakis (2001). It is to be stressed the re-appropriation of European elements within contemporary Irish drama operates dynamically, activating fresh responses and discourses, provoking new considerations of hybrid identities.

Among newer writers, two plays that explore more flexible notions of identity focus on football – as Kearney remarks 'by the 1994 World Cup the Irish team would be over half British-born with a centre forward called Cascarino'. (1997, p. 5) The first, Jones' comedy *A Night in November*, was premiered by DubbelJoint at the 1994 West Belfast Festival – toured widely, including Dublin's Tivoli during June 1997. With minimal props – shorts, a World Cup T-shirt, hat and scarf – the piece demands a virtuoso 'monologue' performance of the role of Kenneth McAllister an Ulster dole clerk, creating an impression of his interactions with a range of individuals. Lighting and sound effects create scene change and atmosphere, though a three-level red white and blue rostra (Belfast) is changed to green white and orange when he reaches Dublin airport. Jaded by his marriage, uptight in-laws and aspects of life in Ulster, Kenneth is sickened by sectarian hatred he witnessed on the November night when the Republic of Ireland qualified for the World Cup by beating Northern Ireland – especially evident in his father-in-law's bigotry, 'salt of the earth racism' (1995, p. 17). During this match he subtly helps a Republican supporter to remain safely unrecognised. Intrinsic to the piece, but not overly didactic, is his growing sympathy for and awareness of both the relative freedom in his (unusually) Catholic colleague Jerry's lifestyle, and a range of disadvantages experienced by the 'Fenian' community. The paradox of Irish identity is epitomised within the teams:

> [...] while this man who was born in England, was playing football for the Republic, in Belfast he was called a dirty Fenian bastard [...] a player born in Belfast playing for Northern Ireland was also called a dirty Fenian Irish bastard [...]
>
> (ibid., p. 17)

After speaking out at a social gathering during April, at the start of Act Two, Kenneth is told 'You are British and should be ashamed of yourself', for his questioning of the right to hate (p. 35). Instead of joining the golf club (for Protestants only), he crosses the Border for the first time in his life, using the money to fly to New York from Dublin with Republican team supporters for their match against Italy. Not only does he experience the cameraderie of these supporters, but he finds other clandestine men from the North joining in 'Jackie's army'. Exhilarating physical performance captured, without descending into sentimentality, Kenneth's carnivalesque experience of flight and bar as 'wall to wall Irish men in green, white and gold'. Despite failing to get a ticket, Kenneth's old values are finally overturned in the face of the welcoming comradeship he finds. Attributing this to the history of migration he muses;

> ... from that day it was the unspoken rule that the Irish would have to look after their own ... even me, who never considered himself an Irishman ... in their eyes I was one of them and I loved it.
>
> (p. 43)

Girls in the bar mouth the Irish National Anthem's words to help him – as he had helped the Republican supporter sing the Orange anthem, 'The Sash', back in Belfast. While victory over the Italians is celebrated, news comes of a shooting of men watching the match, in a pub near Belfast. Kenneth finds it impossible to connect this atrocity with the warmth and celebration of humanity around him, finally disassociating himself from his past he cries out:

> ... tonight I absolve myself ... I am free of them ... I am free of it, I am a free man ... I am a Protestant man ... I am an Irish man.
>
> (p. 47)

At this point in Dublin there was a long, standing ovation for this performance of a hybrid identity.

Bolger's *In High Germany* first staged at the Gate Theatre, was part of the 1990 Dublin Theatre Festival. Set on platform four of Altona railway station in Hamburg just after midnight on June 19th 1988, it too is a monologue, spoken by Eoin, centring on a match – one the Irish Republic lost to Holland during the European Championships. Eoin is typical of diasporic workers who, despite being called 'the chosen generation free at last to live in your own heritage' (2000, p. 82), are driven by economic and other needs to cross and re-cross European borders in search of work. The play suggests for these scattered individuals, the football terrace where they can evoke the memory of Ireland through the skills of their players, provides a slippery and temporary form of cultural identity, especially in the face of opposing supporters of different origin. It marks a transition from postcolonial anti-British slogans as a maker of cultural identity to a more hybrid conception (ibid., p. 73). The railway setting – as in *The Lament for Arthur Cleary* – increases the sense of transience. Eoin's talking and projection of other characters, his father's migrations and then final job at an American factory in Ireland sketch aspects of the past. Originally Eoin had disliked foreign-born players in the Irish team as 'they didn't fit in with my vision of Ireland' (p. 88). His European experiences, especially settling down with his newly pregnant German girlfriend, change his attitudes. Feeling 'balanced on the edge of two worlds' (p. 94), his impending fatherhood marks his maturity and the end of football-supporting with his old friends as a means of identity. As Holland wins the match, he realises the Irish team is now in a sense the only country he owns, 'playing for all those generations written out of history.' (p. 97). In a moving speech acknowledging past and future streams of the diaspora, he realises this is, ' [...] the only Ireland I can pass on to the son who will carry my name and features in a foreign land.' (ibid.).

> Thirty thousand of us stood as one on that German terrace, before scattering back towards Ireland and out like a river bursting its banks across a vast continent.
>
> (ibid.)

This vision of unity in plurality is echoed in the final hybrid chant of 'Ole, ole, Ireland'.

Where these two football plays explore the potential for flexibility of identity in a wider context, Lynch's *Pictures of Tomorrow* (1994) reveals a concern for European issues beyond Ireland. Through fragmentation of time it slips between present-day North London and the past Spanish Civil War. Where Ray, Len and Hugo, once members of the International Brigade, are each played by two actors to show the difference/split between their old and younger selves, one actress plays both Kate, Ray's daughter in the present, and Josephina, a young Spanish anti-fascist activist. More radical in form than Lynch's Belfast-centred plays, it suggests both the differing extent of the men's disillusion with their past aggressive class-centred political idealism and ultimately their acceptance of the need for change. Kate wants Ray to enter a care home for the elderly – but he has summoned his old friends on the planned moving-day. From outside, Irish and Caribbean music signifies the community festival – which Ray suspects of having an underlying political agenda. (1996, p. 168) This festival foregrounds the contrast between harder and narrower past ideologies and newer, 'softer,' broad left and representative minority group approaches. How far the memories of Ray (a London Communist), Len (a Welsh miner) and Hugo (a Stalinist from Belfast) are tinged with youthful energy and romanticism, as opposed to their attitudes today, is revealed through juxtaposing past and present, including downbeat details of intervening years and failing health. They recall a poem written by their Irish comrade, Eamon Downey, killed in the struggle against Franco: 'As my spirit spirals forward seeking strength, I catch myself painting pictures of tomorrow' (ibid., pp. 177–8). Act One ends with past and present identities joining in a climactic celebration of their comrades from many nations all 'fighting together for the dignity of man' – undercut by Kate's comment that the war ended years ago. Their real reasons for going to Spain are followed by debate about changes in Eastern Europe. Hugo leaves, feeling let down by Ray's conviction that communism is dead, and by Len's assertion of the need to change. Momentarily Ray feels all his life has been wasted, despite his past dream of socialism:

> We're living at the end of hope. When we were young, we could dream great dreams. Now what have the kids got? Materialism and nationalism. Bosnia. Northern Ireland. We have to sit and watch narrow-minded men fight over lines on the map.
>
> (p. 221)

Inspired by a vision of Josephina leading a roll-call of the International Brigade dead, Ray decides to leave for the home, wearing his Civil War hat, quoting from a Ulster poet,

> I feel like a green shoot, waiting for a flower.
>
> (p. 223)

Not only this sense of the possibility of change, but seeing the Troubles within a European arena where other nationalisms are in conflict, provides an optimistic, outward-looking context evident here and elsewhere. In considering international politics further in terms of American neo-imperialism, and especially other nations' troubles in the Middle East, McGuinness' *Someone Who'll Watch Over Me* (1992) explores a hostage scenario. Reminiscent of Sartre's *In Camera* (1944), not to mention the real-life incarceration of John McCarthy and Brian Keenan, it is set in a Beirut prison, in which Adam, an American, dies. Consequently, an Englishman and an Irishman are compelled to form an understanding, despite their previously held stereotypical antagonisms. Much of the comedy depends on reversing such stereotypes – but the widening of dramatic contexts beyond the postcolonial is the most significant element. However, McGuinness' *Dolly West's Kitchen* (1999), which embraces American, European, English and Irish identities from both sides of the border, goes further towards the possibility of reconciling a 'postcolonial cosmopolitan internationality with the ties of tradition and locale' (Scott Brewster, 1999, p. 127).

Opening at the Abbey, *Dolly West's Kitchen* was seen at London's Old Vic, early in 2000. Although events take place when the Republic was neutral during World War II, in Buncrana County Donegal near the border with Derry, the West family's experiences have post-millennium resonance, especially through the generally positive image of future, hybridised multiple cultural identities. In London, the set – though without the enclosing three walls – showed Dolly West's kitchen in relatively realistic detail, emphasising the central table, a typical trope of Irish domestic settings. Significantly it suggested a surrounding liminal space, which through changes of light and sound was transformed into a fuller impression of shore/seascape for Act Two, Scene Two.

Throughout, the sound of the sea is heard at key moments. This play re-works several traditional themes, including migration and return, but emphasises breaking down barriers and prejudices.

Events hinge on the way Rima, a sixty-year-old mother, eliptically sets up certain tensions, which act as a catalyst for change in the lives of her three children, Esther, Dolly and the much younger Julian, born after her philandering husband returned home to die after ten years away. Rima, who 'believes in this world, not the next' (1999, p. 63), is unconventional and sometimes outrageously forthright, but apart from tiredness shows no signs of her imminent death. It becomes clear she understands the underlying reasons for her family's turmoil, and is accepting of difference whether of nation or gender in welcoming visitors to the house, joking even about the attitudes of Protestants and Catholics. Dolly, her younger daughter who left her own restaurant in Italy closed for the duration of the war, has taken over the cooking at home. She has long loved Alec, a possibly bisexual Hibernophile, now based as an English soldier just over the border in Derry. Esther has an unsatisfactory and barren marriage to rather passive Ned, who allows her to do what she wants. Julian is a repressed, narrow-minded individual who wants to ban Alec from coming into the house, because although he is an old friend, he is a representative of a colonising power. Like Ned, Julian is a member of the Irish Republic's army of defence. Rima deliberately introduces into the house Marco and Jamie, two American GIs based in Northern Ireland, whom she met in a pub when they were in the Republic for rest and recreation. As Rima admits to Dolly, she has thus brought into the house, 'Badness. Good isn't it? A bit of badness' (ibid., p. 30). Marco, who by his overt admission at first behaves like a 'Twisted mean sissy queen' (p. 61), gradually erodes Julian's defensive bigotry and they fall in love, though it is only when Marco manages to recover from his war experiences that they are able to leave together for Italy after the peace. Jamie's affair with Esther eventually provokes Ned to fight for his wife, and in the last scene Ned is seen pushing a pram around the edge of the sea/house with his baby daughter – whose father remains ambiguous despite Jamie's denial that she is his. Esther, staying in the family home with Ned, forgoes her love for Jamie, who is snapped up by the West's orphan maid Annie, keen to marry him and live in America. Dolly finally and reluctantly agrees to go to live in England with Alec.

Performance makes it clear that personal movements towards and away from Ireland were linked both to history and the wider world, just as elements of theme and staging transcended the apparently realist form. Flexible relationships across borders, whether Republic/Northern Ireland/England/Europe/America, or across gender, masculine/feminine/bisexual/homosexual, were shown to be liberating and thus desirable. Stereotypes such as Madonna/Magdalen/Matriarch or Playboy/Cowboy were subverted. Rima also shows an ironic awareness of those new authenticities, which circulate through intertextual use of cliches of images and songs about Ireland, such as 'a little bit of heaven fell out of the skies one day' (ibid., p. 32). Her last speech about being a seabird, flying over the whole earth, before she returned to touch the soil of Ireland in her own garden is reminiscent of Maeterlinck's *Bluebird* (1908). Whereas the kitchen table remained the focus for feeding both relationships and family tensions, the shore – the borderline between land and sea – was a heterotopic site for transformation, rather as in Jones' *Women on the Verge of HRT*. After Rima's funeral it is on the shore that key interactions triggered by her take place. Marco and Julian confess their love, Dolly confronts Ned and Esther to make them face up to their marital crisis, she and Alec at last confess their mutual feelings, and Anna finally succeeds in seducing Jamie. In the kitchen, the last family meal – symbolically bread and wine – is served before most of the group disperse to various destinations, although Ned still looks inside from the garden. Outside Alec and Dolly speak different verses of the hymn 'I vow to thee my country', with their different homes of origin in mind. As the others inside clink glasses to Julian's quotation from the marriage service, the play ends with Alec's 'Is the war over?' and Dolly's reply 'I said I hope so' (p. 85). Of course, as the play is set before the escalation of the Troubles in the late 1960s/early 1970s, there is an irony behind these lines, which for audiences in 2001 may evoke the present faltering Peace Process and its problems. Nevertheless this mediational modernist play celebrates the possibility of multiple, hybrid identities as well as a broader definition of family relationships, which may still endure despite geographical distance, and looks beyond the limitations of a postcolonial nationalism.

In contrast, set in the present, Gary Mitchell's award-winning *The Force of Change* (2000) examines the prospect of political change in Northern Ireland but uses an intensely claustrophobic form of realism to interrogate – in thriller format – the nature of the Royal Ulster Constabulary, and of the now-banned UDA. Aware even of the actor's breathing, the Royal Court Upstairs audience sat on either side of a relatively small traverse performance space, set to suggest two police station interview rooms which were almost mirror images of each other. Caroline, married with two children, is a Detective Sergeant due for promotion – an unseen Inspector is keeping a record of her work. Although her work partner, Mark, is relatively co-operative and supportive, clearly she has to battle against her male colleagues' deep-rooted sexism. That any Catholic colleagues now in the RUC face similar prejudices is implied throughout. Caroline is being stone-walled by the silence of Stanley Brown, held for a limited period on suspicion that he has coerced witnesses relevant to his involvement in the UDA killing of a Catholic pornographer into changing their evidence. Also under investigation is Rabbit, a younger joyrider, whom the police hope to press into confessing he steals cars for Stanley and thus the UDA. On stage Stanley's silent presence exudes menace, in contrast to Rabbit's almost comic confusion, but the play reveals the difference between taped and 'off the record' police behaviour. Bill Byrne, an older Detective Constable, jealous of Caroline's career, dislikes her more inventive interrogation techniques. As Act One ends, in Caroline's momentary absence, Bill agrees to Stanley's request for her personal and car details. This dangerous betrayal is intensified in Act Two, when Stanley is found to have Bill's mobile phone. Bill has often done small 'favours' for the UDA, who had bought out his debts. Rather than inform the Inspector, the men want to deal with Bill themselves. David, a younger Detective Constable, reveals his prejudice against women, Catholics and those who consider the RUC to be corrupt and bigoted. He condemns RUC leaders as useless neutrals who have 'turned us away from our traditional enemies to block, fight and hurt the people of our own communities' (2000, p. 67). In supporting Bill, he attempts to get at Stanley by implying that the RUC secretly still has the same objectives as the UDA. Meanwhile

Caroline manages to 'break' Rabbit by threatening they will release him to be dealt with by some tough UDA supporters, but unfortunately he incriminates another individual, not Stanley. Sticking to her principles, she refuses to allow Mark to beat Rabbit into naming Stanley. Stanley does not respond to David's threats, so ultimately he is released.

David's remark, 'How safe do you feel now?' as the play closes addresses not only Stanley, but by implication the audience, positioned in the original production as the restricting walls of the interview rooms. Within this literally enclosing space there seems no possibility of heterotopic transformation. Despite its pessimism, the play does reveal dawning processes of change not only in the increasing involvement of women – but also by implication, Catholics – within state institutions. Further, despite the disorientation of the RUC personnel, it provides a powerful critique of that part of the Protestant community which turns a blind eye towards criminal activities by terrorist sub-groups as, seeing 'the Nationalist community prosper whilst they lose jobs, prospects and basic civil liberties' (2000, p. 76), they long for a return of a 'heroic' struggle. David points out to Stanley that no one outside North Belfast continues to see the world through such falsely romantic perspectives, and

> [...] when I look into your eyes I see a reflection of every IRA man you claim to be protecting us from.
>
> (ibid.)

Thus, while this play terrifyingly represents the impact of political changes, it challenges old prejudices about cultural identity through the kind of parity of disesteem evoked by Coulter (1999, p. 257) – thus heralding an era when it will be possible to look beyond the past's fixed perspectives.

Paradoxically, Mitchell's use of realism is more challenging than certain plays that disappoint ideologically while deploying some potentially radical strategies. Michael Harding's *Hubert Murray's Widow* (1993), set in Fermanagh, 'should not be presented as realism, but more akin to a dream' (1996, p. 4). A complex thriller-style plot is touched with black humour and flashbacks, depending upon the presence of revenants, an IRA man and his widow's Protestant lover. Concentrating on Phaedra-like passion, it is cynical about Ulster:

> The problem has no solution. Violence is just a way of imposing order on chaos. The problem itself ... is the solution.
>
> (ibid., p. 42)

Similarly Harding's *Misogynist* (1994, Peacock), with 'an austere no man's land' cluttered with Catholic iconography, video screens and a chorus of monks finally revealed as women, seems rather unconvincing in exploring gender identity, despite contrasting verbal dominance of an older male with a younger female's potentially subversive physicality.

Contemporary developments in drama that successfully transcend both the forms and ideas of the past are reflected in a proliferation of groups evident in programmes for large annual festivals, such as those in Dublin, Belfast and Galway, and performances in other towns, including rural tours to small non-theatre venues. It is impossible to document all this activity, not least because, despite some use of video recording, only relatively few new plays gain wide – if any – publication. New Island/Nick Hern, The Abbey/Methuen, The Abbey/Syracuse, Rough Magic and Faber have produced anthologies cited in the Bibliography, while Druid/Royal Court, Nick Hern, Gallery Press and Lagan books have published single theatrescripts. Despite difficulties around the need to link production with short-run publication, as Peggy Butcher of Faber has pointed out (*Irish Review,* 1998, pp. 9–15), there is a need for wider access to publication for new work. A central reference-library style comprehensive 'bank' for new Irish scripts could be established, similar to that for Women's Writing at Bristol University, begun by Linda Fitzsimmons. As researched by Belfast-based Dr Anna Cutler, physical/visual theatre provides an even more difficult challenge for documentation.

A brief account of some key groups from the Republic and Northern Ireland respectively, based on field work, illustrates the growing tendency to look both beyond national concerns and the restriction of verbal theatrical forms, thus fostering flexible notions of cultural identity. In Dublin in 1997, the Abbey's production of Patrick Kavanagh's novel *Tarry Flynn,* which visited London's National Theatre in 1998, was a highly physical piece stylistically suggestive of Jacques Lecoq and the UK's Théâtre de Complicité. Originating in different localities, all companies cited perform across the borders, especially in festivals. The importance of Druid and Rough Magic was suggested in Chapter Four through the careers of Garry Hynes and Lynne Parker respectively. A Druid poster celebrating the twenty-one years from 1975–1996 shows an impressive range of productions, including many plays discussed here, as well as plays from the Irish tradition and European canon such as those by Brecht, Buchner, Dario Fo, Ibsen, Shakespeare and Greek classics. In June 1997, the administrative staff consisted of three full-timers, the company had funding from the Arts Council as well as Local Authority Funding, and their planning was essentially project-based. In 1996 they had received a special touring grant for the islands off the West Coast, but were then looking forward to their 1999 Synge Festival. Funding does allow some support for commissioning writers – Vincent Woods and Meliosa Stafford have been Writers-in-Residence – but the huge amount of unsolicited scripts, is vetted by a series of readers, then a panel. Although the visual/physical impact of Druid's repertoire is very different from groups such as Barabbas, Druid has often staged Irish canonical works in a contemporary way – for example, Mustapha Matura's engaging re-working of Synge in *The Playboy of the West Indies,* in association with London's Tricycle in 1994. The award-winning *Leenane Trilogy* was shown in their Druid Lane venue, and Galway Town Hall Theatre is sometimes used, but Irish tours and their collaborative ventures with London theatres such as the Tricycle and especially the Royal Court, which have led on to America, indicate an ever-widening and international audience.

Rough Magic's great strength is the core of performer/writers who have worked together from the first, a collective experience acknowledged by Parker in their Anthology's Introduction, and evident in their responsibility for many plays discussed earlier. The group, which now has considerable experience of international touring, began by producing a wide range of contemporary and re-worked classical plays, but new writers are central to their policy. The work of Declan Hughes, their Writer-in-Residence, discussed in Chapter Six, is especially relevant to postnationalist concerns. On the other hand, Paula Meehan's *Mrs Sweeney* (1997) is at one level a sharp critique of life's harsh realities in Maria Goretti Mansions, a rundown Dublin council estate housing block, at another an exploration of female relationships and also a carnivalesque transformation drawn from a traditional myth. Lil Sweeney's flat is burgled yet again as the play opens. She still talks to her daughter, Chrissie, a drug addict who recently died of AIDS – not unusual for those brought up in this deathly environment. Lil's life-long friend, Frano, a battered wife, and Mariah, Chrissie's friend, who is struggling to keep straight in the hope of a job running the Women's Project for the Community Development group, both provide emotional support and drinking companionship for Lil. Sweeney's breakdown, the play's crux, is triggered by local vandals destroying his beloved pigeons. Father Tom, a well-intentioned but ineffectual Catholic priest and Oweny, Sweeney's garrulous friend, attempt to be supportive, but Lil refuses to have her husband hospitalised. By Act Two Sweeney's stunned silence is replaced by his fantasy that he is a pigeon; fluttering and making bird noises he nests in fragmented newspaper under the table. As the women prepare for an All Soul's Night community carnival by donning costumes as exotic birds, he is transformed into a magnificent peacock. Lil, who has told the bleak fortunes of the others but cannot see her own, ritually toasts her dead daughter, evoking the swirling flights of pigeons in a speech that merges past, present and future, ending with Chrissie's last word, 'Magic' (1999, p. 462). Earlier noisy eruptions of vandals outside are replaced by the women's singing inside and then the band outside; growing darkness and sense of entombment in the boarded-up flat contrast sharply with the bright carnival materials. Despite surrounding squalor and tragedy, the three women's courage provides a moment pregnant with transformative possibility. This physically demanding re-working of *Suibne Geilt* echoes the hero's descent into madness after experiencing the horrors of the battlefield (Smyth, 1996, pp. 158–162), and is a typical postcolonial strategy.

Donal O'Kelly, the founder of Calypso theatre company has written and performed with Rough Magic, often working in a surreal intensely physical mode, which challenges notions of identity. A director of justice and peace organisation Action from Ireland (AfrI), he has outward-directed political concerns, also echoed in Calypso's 1995 Manifesto as 'changing the world – in small significant and possible ways' (*Irish Review*, No. 1998, p. 28). *The Business of Blood*, co-written with Kenneth Glenaan, is described by Vic Merriman as an attempt at 'cultural re-mapping', which places the North/South relations within a wider geopolitical context (ibid., p. 29). O'Kelly's third solo play, *Catalpa*, also seen at the Gate, Dublin, won an Edinburgh Fringe First Award in 1996, going on to London and Melbourne. *Bat the Father and Rabbit the Son* (1988) premiered by Rough Magic as part of the Dublin Theatre Festival, then toured Britain, Ireland, New York, Australia and New Zealand. Performed solo, it shows inter-generational tension between the successful international haulier Rabbit and his dead father, Bat, who as a revenant erupts into his son's body and speech from time to time while he is talking to Keogh an invisible assistant. Middle-aged Rabbit – indicative of growing material prosperity – somehow cannot quite come to terms with memories of his gentler father, a pawnbroker's assistant and veteran of the Irish Citizen Army of 1916. Events include childhood fishing and a sort of odyssey round Dublin Bay in search of a green glass buoy, into which fragments of the past are woven. Rabbit, despite his attempt to reduce all values merely to the cash nexus, fails to obliterate the voice of his 'failure' father and dies. *The Dogs* (1992, Dublin Theatre Festival), O'Kelly's first non-solo play, is even more disruptive of realism in its subversion of a Dublin domestic setting, where usual family Christmas tensions are to be exacerbated by a pregnant daughter bringing home her partner from Ulster for the first time. Each Rough Magic actor also performed an animal role – dogs Pax and Rebel, the Turkey, a Rat and a Robin – which in some way reflects their human persona. The dogs mostly live in a limbo-land – the surreal world of undead animals – a borderline situation from which they want to escape, but they and the turkey are transformed into swans in a miraculous ending. In animal scenes – and at points where humans and animals interact – strange scrabbled Joycean language, often touched with scatalogical references and historical references to Irish political crises, is used. Grotesque and carnivalesque touches reverse and challenge traditional religious and political terms and values. O'Kelly's Afterword compares the messed-up turkey to the chaos of the Troubles, and suggests signs of the Peace Process are indicative of forthcoming compromises and peaceful solutions, like those made by the family towards the end of the play. Overt references to the role of Europe (1999, p. 142), the Troubles and the possibility of change (ibid., pp. 173–4) look forward to the future. Despite the serious subtext, physical energy must fuel absurd elements to comic effect.

O'Kelly's *Asylum! Asylum!* (1994), premiered and published by the Abbey, is written in realist form but explores the increasing topical issue of asylum seekers. Whereas in 1992 a total of 80 individuals applied for asylum in Ireland, in early 2001 the rate was reported as 2,000 applications per month (*The Observer*, 18 February 2001, pp. 16–18). Despite support for Third World Aid typified by ex-President Mary Robinson's policies, this article and a previous one by Tom Paulin have discussed the extent of unacknowledged racism in Ireland (*The Guardian* 29/08/2,000). This intensely moving play derives from two real events, the burning of prisoners in a log-covered pit in Northern Uganda and the torching of a Vietnamese immigrants' hostel in Germany. It explores harsh attitudes shown by Dublin's Immigration Officers in implementing EEC policy:

> It only takes one leaky section in the walls of Fortress Europe and the flood of immigrants will pour in and swamp the continent.
>
> (1996, p. 153)

> [...] the only way to get zero is to instil fear as a deterrent.
>
> (ibid., p. 165)

Through contrasting go-getter officer Leo and his sister Mary, a liberal-minded solicitor to the case of Joseph, an asylum seeker, the play exposes parallel relationships between fathers and children endemic to this moral dilemma. Joseph had been tortured into being complicit in a prisoner-burning episode – like the one above – in which his father died. Different attitudes foreground strategies for turning refugees away and the hypocrisy of those who applaud them:

> We're marking them! We're impounding them in camps! We're forcibly transporting them! We're calling it a solution! [...] It happened once before [...] Dad remembers. It's happening again.
>
> (p. 168)

Expressing Leo's road-to-Damascus-like change of attitude, these lines are a challenge not only to Ireland, but to the rest of the EEC – and resonate through the family's determination to stand up for the rights of asylum seekers, despite the unjust transportation of Joseph.

Passion Machine produced twenty nine new plays between 1984 and 2000, and continues under the inspired Artistic Directorship of Paul Mercier to

> create and promote a wholly indigenous populist theatre that depicts, challenges and celebrates the contemporary Irish experience.
>
> (Policy)

Mercier, an ex-teacher of extraordinary energy, has among these directed 10 of his own plays, while his *Down the Line*, commissioned by the Abbey, was directed by Lynne Parker at the Peacock from late September 2000. Drawing upon a community ethos, the bulk of Passion Machine's productions have first been staged at Saint Francis Xavier Hall in Northside Dublin (SFX), while the Office and rehearsal rooms where Mercier gave a generously long interview in 1997, are more central. The group has also toured both sides of the Irish border, to London, Poland and Greece. Writers involved have included Roddy Doyle, Gerard Stembridge, Michael Harding, Nell McCafferty, Anto Nolan and Declan Lynch, while annual funding has come from the Arts Council of Ireland, from Friends of the Company and sponsorship from industries such as Yamanouchi due to the patronage of the Chief Executive. Mercier is especially concerned to challenge attitudes that claim 'the portrayal of ordinary people is not as important as spiritual extremes of experience just a little bit remote from reality' since he rightly believes that '"living" theatre is not less than [...] professional mainstage theatre' (Interview). *Studs*, winning a Fringe First Award in Edinburgh in 1992 and the Rooney Prize for Irish Literature, then made into a feature film, is typical of Mercier's belief that 'Youth is constrained to look at the here and now [...] Drama has to have that kind of energy' (Interview). Mercier's *Buddleia* in its original 1995 production crammed 29 boisterous actors into the small space of Project Arts Centre, and climaxed with destroying the set! The re-staging of his full *Dublin Trilogy* (*Native City, Buddleia, Kitchen Sink*), which critically celebrates the inhabitant's experiences of socio-economic changes, received the 1998 Dublin Theatre Festival's Best Production Award.

Mercier's own plays and those he directs embrace different forms and styles. Declan Lynch's *Massive Damages* (1997) seen at the Tivoli, Dublin, used Brechtian, comic techniques to satirise the law, the media and show business through two anonymous alcoholics who defraud the system and destroy reputations. *Kitchensink* (1997) explores problems caused by housing sprawl through which greed for profit destroys the environment, through two generations of a family in such a house – but uses masks to suggest an allegorical tale with ritualistic Greek dimensions. Just as the river in this play is built over, so Mercier suggests that the spiritual is forced underground – 'We are loosing touch with the agents that shaped the landscape [...] we are loosing our sense of place'. Further, the theatre's function is to enable us to journey, ' [...] discovering new frontiers, not geographical frontiers but frontiers in the imagination' (Interview). *Down the Line* (2000), despite having an apparently realistic fourth-wall set comprising a suburban Dublin home with neat kitchen/cluttered garage and implied garden, condenses 10 years of the Walsh family's life in the 1980s into a one-day time span, which also gives the four seasons. Set changes are organic to the process of scenes, so that bending time produces 'six contradictory snapshots of the family'. Mother and Father who suffer mutual communication problems have four children who each represent changes – especially now the young may choose to migrate rather than be forced to do so (2000, pp. 114–5). Sean, who works in London, marries pregnant Barbara and returns to Ireland, but later divorces her, taking as his partner Marie who is working 'on a feature on Irish ex-pats in the context of [...] this big thaw' (ibid., p. 119). Dierdre, the youngest, starts as a radical youth/social worker, lives with her boyfriend, but later returns to London having had a breakdown due to overwork. Liam, a rock musician, has children with a New Age girl, while Maeve the older family-orientated daughter marries a chef. Mother's rural roots and her later taking up work with a solicitor, like Father's gardening as a contrast with modern materialism, all suggest change is double-edged.

Native City follows three Dublin families across the centuries in flashback, starting in modern times with a racist attack on two inoffensive Bosnian refugees, thus linking both inward and outward perspectives on history. The plight of migrants in both directions is also seen in Mercier's *We Ourselves* (May 2000) and Joe O' Bryne's *It Come Up Sun* (November 2000). In the former, seven monologues are delivered by young Irish individuals, who met when working in a German factory, moved to Eindhoven and then separated. Starting with Sarah, a woman politicised by her multicultural encounters in Germany, and ending with Declan, an Irish French-speaking Eurocrat, spending a night in Dublin before returning to his Brussels home, each monologue not only reveals a character at a crisis point but covers critically socio-political and cultural trends of the 30–year evolution from Emerald Isle to Celtic Tiger. Jocelyn Clarke suggests (*Irish News*, 2 June 2000) that the diversity of Irish experience and 'a world view which is inclusive, progressive and above all pluralist' is especially evident in this contrast between Sarah and Declan, not to mention a gay man determined to baton-twirl in New York's Saint Patrick's Day Parade. The *Sunday Tribune* comments that Mercier's topic is urgent but seldom addressed in drama,

How Irish identity is being challenged and transformed by the influx of new arrivals and by the country's increasing Europeanisation.

(21 May 2000)

O'Byrne's play, directed by Mercier, is set in Dublin's dockland amid cargo-containers. Billy, a paranoid schizophrenic with manic but harmless tendencies has become friends with stoical Joe, an unhappily married nightwatchman. Ania, a young illegal immigrant from Eastern Europe, is found in a container with the body of her dead mother. Questioning the nature of loneliness and the equalities of the human condition shared by all three protagonists, the play transcends national boundaries.

The scope of Pan Pan theatre, which works with deaf and hearing actors, has always extended beyond Ireland, as evident in its International Theatre Symposium, first held in the Samuel Beckett Centre, Trinity College Dublin in 1997, with its fourth in January 2001. Founded by Gavin Quinn and Aedin Cosgrove in 1993, it has produced 12 original pieces in a variety of styles and content. Tours have included the UK and Ireland as well as festivals in Poland, Italy, France, Austria, Sweden, Denmark, Korea and Australia, with the intensely physical *A Bronze Twist of Your Serpent Muscles* winning the Dublin Festival Best Overall Production Fringe Award in 1995. The Symposium, attended by 'cutting edge' groups, aims to exchange contemporary approaches across Europe and encourage 'debate between International theatre companies and Irish audiences about the development and the changing role of theatre in contemporary society' (Pan Pan leaflet, 2000). Typical in its exploration of the different languages of theatre, *Tailor's Requiem* (1996) drew upon dance, mime, music and performance art to produce vivid theatrical images, also evident in other works such as *Cartoon* (1998–9). A later collaboration with Dermot Healy created *Mr Staines* (1999), which Quinn describes as sitting 'between content and 'thoughts of a spectacle' resting simultaneously in the idea of memory and recollection' (programme note). *Standoffish*, which had won an Adelaide Fringe Award on touring abroad before showing at the Dublin Festival in October 2000, is ostensibly about three diary-keepers. It carried even further their post-Artaudian strategies by breaking audience/spectator barriers, not only through hugs but also handing out a bloody heart. Their body-centred approach included onstage peeing amidst manically fast physical movement, deconstructive grimacing, song and flashing lights, in the kind of experimental performance that can never be experientially recaptured.

Barabbas (Dublin) and Macnas (Galway) are also companies with an intensely physical emphasis beyond fourth wall realism. The former, formed by Veronica Colman, Raymond Keane and Mikel Murfi in 1993, is primarily visual, physical and dynamic, following 'the traditions of Commedia dell'arte, Clown and Buffon to create a style of Irish theatre from these previously European-dominated traditions' (programme notes, *Sick Dying Dead Buried Out*). Their range extends from 'non-verbal, lyrically visual devised productions to commissioned work and new interpretations of extant texts' (programme notes, *Gods Gift*); from Shakespeare's *Macbeth* to *Half Eight Mass on a Tuesday* (both 1993) or *Out the Back Door* (1997), a children's show at The Ark, Dublin. The company has conducted physical theatre workshops at home and abroad in London,

Wales, Denmark, France, the USA and New Zealand. An especial triumph was their reworking of Lennox Robinson's *The Whiteheaded Boy* (1997), with one actor playing the 'boy' and the three founding members playing a multitude of roles across gender under Gerard Stembridge's direction, deploying the original stage directions, in tongue-in-cheek unusual ways. Currently funded through multi-annual funding with the Irish Arts Council for three years, Barabbas receives around £2,000 annually from Dublin Corporation, small amounts from the British Council for workshop tours of the UK, with further support from the Goethe Institute in 2000 for John Banville's adaptation of Henrick Von Kleist's *Amphitryon/Gods Gift*, which featured in the main Dublin Festival programme and also toured. This version mixed Roman comedy with a chaotic Irish context. Earlier in 2000, *The Cube* – a self-propelling 8 x 8 foot shape moved through Meeting House Square at the speed of one mile per hour. An audience of 24 people, perched on its outsides, fitted with headphones to perceive an eight-minute animated performance within. Small wonder this innovative ensemble has earned an Entertainment Award in the 1997 Pan-Celtic Film and TV Awards, three nominations for the 1998 ESB *Irish Times* Theatre Awards, and the 1998 Kilkenny Beer Cream of Irish Theatre Award for their contribution to Irish Theatre. Mikel Murfi, who trained with the influential École Jacques Lecoq school in Paris, has also worked with Macnas as a freelance director.

Macnas, established in 1988, is at the forefront of community arts and cultural activities nationally and internationally. It had an extraordinary office/workrooms site in the Fisheries Field on the edge of the River in Galway, when Company Manager Declan Gibbons was interviewed in 1998. At that time Macnas was funded annually by the Irish Arts Council and also Integra, European Social Funding from Brussels to provide two-year training in Community Arts to combat social exclusion. Gibbons strongly felt,

> Everyone has a right to participate in the creative experience [...] in the Macnas context everyone can contribute.
>
> (Interview)

Accreditation for a Higher National Certificate in Public and Community Arts was to be developed for a target group of long-term unemployed people in certain areas of Galway city, involving a project production, *Message in a Box,* and including trainer exchange with the Paroles project in France. Aiming for immediate context and accessibility, Macnas' performance work has been carnivalesque in style – for example the huge street *Carnival of Fools* central to the 1998 Galway Festival, involving masks, gigantic figures and all the reversals of misrule and body-centred grotesqueries described by Bakhtin ((ed.) Morris, 1994). Influenced by European elements, including Footsbarn Theatre Company from France, the company's style is visual and physical, often including masks and puppetry – to combat, as Gibbons suggested, the more verbal text-based traditions of Irish theatre, even when devising versions of cultural myths such as *Balor, The Tain,* or *Sweeney.* This French influence was evident in the highly praised, intensely visual *Diamonds in the Soil* (1997) about Van Gogh, devised and designed by Patrick O'Reilly and directed by Mikel

Murfi. Fluid movement, haunting music and 'coming to life' of the artist's pictures embodied his mind's deterioration, brilliantly performed by French-speaking Antonio Gil-Martinez. Macnas' adaptation of Patrick McCabe's novel *The Dead School* (1996), seen at the Town Hall Theatre in 1998, was also postmodern in its fragmentation and stylised physically adept ensemble performance. During 2000 the company moved into multi-annual three-year funding with the Irish Arts Council. Originally shown at the Galway Festival, *The Lost Days of Ollie Deasey* won the Best Irish Theatre Award at the Dublin Festival in 2000. A re-working of Homer's *Odyssey*, hero Ollie, a hurling champion was presumed drowned years before. His son Terry, also a hurler, hears news his father may be alive. Murfi superimposed the shape of Odysseus' travels on a map of Ireland to create Terry's bus journey to find him. Staged in the Round Room at the Mansion House in Dublin (and Leisureland in Galway), actors and audience stand/walk in the same space, with the white lines indicating the playing area. The evocation of a long, rural bus journey was cleverly conveyed by items whizzing past, and finally the area became a sporting arena in which an All Ireland Hurling Final was reconstructed. This post-Artaudian total theatre was critically acclaimed as innovative and exhilarating.

Belfast-based field work suggests companies in the North have concerns similar to those in the Republic, such as international links or community orientation and physical/visual experiments, which challenge both realism and the verbal text. The importance of Charabanc and DubbelJoint has been indicated through the account of Marie Jones' work in Chapter Four. David Grant, of the Lyric Belfast, cited Jones as the first to examine Left/Right politics as opposed to sectarianism (Interview, 1997). As with Martin Lynch, an honourable history of popular, community-based projects has until recently tended to be obscured by the kind of elitist prejudice cited by Mercier (quoted earlier in this chapter), even in an area where they are invaluable. A magnificent, moving example of this work was Dock Ward's production *Rebellion* (1997), about Henry Joy McCracken. A radical Presbyterian in 1790s Belfast, and a founder of the Society of United Irishmen, he was concerned to establish a 'brotherhood of affection' across religious views and involved in forging an alliance between Catholic Defender organisations and the United Irishmen. He was hanged in the rebellion of 1798. Writers, including Lynch, had given advice on the script, the production team had all had professional experience, and Director Paddy McCoey had also been involved as a Development worker for the Community Arts Forum. A large cast across a wide range of age and background performed convincingly on a proscenium stage at Saint Kevin's Hall. Simple use of wooden planks, ramps and effective lighting supported the partly stylised, partly realist performance in which didactic elements in the script were counterbalanced by song. Women interviewed claimed involvement with the group had changed their lives, not only in terms of friendships formed, which made them think beyond apparent differences between people, but also in growing self-confidence. Since the group's first production, *The Dock Ward Story* (1991) sponsorships gained include the Northern Ireland Arts Council, Belfast City Council, the Community Arts Development fund and the European Partnership Board. Formed in 1994 on a smaller scale is Shankill Community Theatre Company, whose production *Saved by the Bell in the Last Round* (November, 1997) at the Crescent Arts Centre, combining music and a wedding invitation, was populist performance celebratory of the ups and downs of local everyday life, flourishing despite old political tensions. In contrast, events such as the Ethnic Minorities Community Festival in the same period, in connection with the European Year Against Racism, suggested a movement beyond these concerns.

Under Leon Rubin in the 1980s and Robin Midgely in the 1990s, the Lyric Theatre Belfast was claimed by David Grant (1997) to be at that time the only full-time producing theatre in Northern Ireland. It had remained open in the evenings during the Troubles, serving mainstream, school and community needs. The theatre also receives productions, for example from Red Kettle, Waterford's *The Salvage Co.* Feeling that new writing tended to be researched and developed by new companies – as a main brief for Tinderbox especially – Grant listed successful Northern writers, from Stewart Parker to Gary Mitchell. He felt much work in the North still drew largely upon verbal traditions, despite the increasing influence of French styles such as Lecoq, or directors such as Gerard Stembridge, whom he described as 'the Irish Mike Leigh'. The Lyric received three nominations for ESB Awards for performances in 1998. Perhaps the lure of Dublin for young Northern writers Grant mentioned has since diminished somewhat due to the Peace Process. As someone who has worked on both sides of the border, he stressed the importance of community work with non-theatre goers, saying

> [...] cultural life was essential. In the long term we have to see a modern cultural identity taking over the old allegiances.
>
> (Interview)

The evolution of the annual Belfast Festival at Queens, sponsored by the City, Guinness and others illustrates this possibility through its central staging of major international events – such as a Robert Wilson premiere at the new Waterfront venue in 1997 – with the very best of Ulster's own talent alongside other visiting performers, artists and musicians.

Kabosh, formed in 1994 aims 'to produce innovative and cutting edge physical and visual theatre' (Mission Statement, 2000). Following restructuring through its three-year Development Plan in 1997, a more ambitious four-year plan now extends their touring schedule beyond Ireland and the UK into Europe. Currently the highest funded Independent Theatre Company Client of the City Council, it has won a raft of Awards: Bass Island Arts in 1999, Granada TV Best Play 1999, Dublin Fringe/*Sunday Times* for best Actor and best Production for *Mojo-Mickybo* (1999). With 14 characters brilliantly performed by only two actors, Owen McCafferty's play charts the cross-sectarian friendship of two boys obsessed by Butch Cassidy and the Sundance Kid. Lottery-supported for fee-based expenditure for new writing and by the Arts Council of Northern Ireland, Kabosh aims for excellence whilst pushing forward performance frontiers. Artistic Director Karl Wallace has implied the postmodern, somewhat filmic, quality of Kabosh style, which 'aims to recreate the whole illusory magic of theatre', has a dangerous and exciting element that can make it more difficult for a company to get started because more parochial attitudes may demand artistic compromise (Interview,

1997). The positive reception of a pacy five-actor all-male version of *Romeo and Juliet* (1999), and *Chair* (2000) – an installation of 29 sofas as a backdrop for a fairy tale, with original soundtrack and precision movement, and providing orange inflated sofas for a reclining audience, suggests that Kabosh has successfully challenged conservatism. *Oriana* showing in February 2001 combines dance, music and physical theatre with minimal use of language in chronicling a little girl lost among a bleak urban landscape.

Included in *Kicking Space* (2000), Kabosh's first Physical Theatre Symposium to which Pan Pan also contributed, were workshops with Ridiculusmus, another award-winning yet small physical theatre company formed in 1992 by Jon Hough and David Woods. Ridiculusmus' shows have achieved something of a comedy cult status both throughout Ireland and internationally, as winners of the Best British Production and Herald Angel Awards in 1999. Seen in the UK and on the Edinburgh Fringe, their *The Exhibitionists*, full of sight gags and slapstick earned, with Pan Pan and Bedrock, the first *Advertiser* Fringe Award at the Adelaide Fringe Festival in 2000. They performed the satirically surreal *Say Nothing* about Kevin, a returning Ulster emigrant, at the Greenwich Docklands International Festival 2000. Their *Yes, Yes, Yes*, seen at the Belfast Festival 2000, was described as 'the dysfunctional offspring of Samuel Beckett' (*The Times*) and 'truly, monumentally, madly silly' (*The Scotsman*). Their playful mnemonic 'ARSEFLOP' echoes their use of chaos as a new art form, which provokes comedy obliquely.

Tinderbox, a touring company founded in 1988, mounts innovative productions of contemporary Irish plays. As the main producer of New Writing in Northern Ireland, its strong commitment is evident in an annual new writing festival and commissioning of writers for full production. Funded by the Arts Council of Northern Ireland, Belfast City Council, the National Lottery and others, the company has won and been nominated for many awards. These include Stewart Parker Awards for Joseph Crilly's *Second-Hand Thunder* (1999) and Carville's *Language Roulette* (1998) – which also won the Arts Council of England Meyer Whitworth Award for Best New Play – and Belfast Arts and Arts & Business Awards. Their repertoire has included plays by the late Stewart Parker, and Jones' *Ruby*, whilst their plan for setting up a Belfast/Montreal playwriting development programme indicates an outward-looking agenda. Their Belfast Festival (2000) production *Convictions*, which took place within the Crumlin Road Courthouse where some of the most notorious Troubles trials took place, has been critically claimed as outstanding. Under the overall artistic direction of Paula McFetridge, seven short plays – by Daragh Carville (*Male Toilets*), Damien Gorman (*Judges Room*), Marie Jones (*Court Room No.2*), Martin Lynch (*Main Hall*), Owen McCafferty (*Court Room No.1*), Nicola McCartney (*Jury Room*) and Gary Mitchell (*Holding Room*) – took place every evening, each in these different spaces and with a separate director. Original music by Neil Martin and various installation pieces by artist Amanda Montgomery intensified the building's damp, decaying and unfriendly horrors. The audience of 80, split into four subgroups, was escorted by 'prison officers' around the performances, culminating in Lynch's *Main Hall*, where the ghost of a previously hanged prisoner laments. Covering a range of perspectives on the Troubles – including disassociation of the middle classes from their reality, and the cynical regret of hacks who have lost a lucrative source of income thanks to the cease-fire – these plays, published in the programme, draw upon both comedy and tragedy. Mitchell's was typically considered the most terrifying, while McCafferty's, in which a prisoner is interrogated by a disembodied female voice, seemed the most experimental. This unrepeatable, innovative site-specific piece was also an important historical and cultural document – Mary Holland (*Irish Times*) has suggested that as this performance could help to liberate both communities from the past, such a repossession would allow 'the building to become a vital part of the city's future'. Susannah Clapp (*Observer* 19 November 2000) more cynically remarks that, as Carville's play implies, there is a danger of 'turning horror into heritage' – an element present in Gorman's Judge who is trying to get audience sponsorship for a Troubles Opera production of *The Ulster War Cycle*! McFetridge, who was involved in the *Wedding Community Play* in 1999, is also a performer who took a key role in the Production Company's *Hunger* by David Brett, directed by Michael Quinn and performed at the Waterfront Hall as part of the 1997 festival. The latter, which treated Belfast as a heterotopic site in depicting a dying hunger-striker, an abused eighteenth-century heroine, a 2000 year old hermit and a make-believe king, deployed a stunning visual style of disturbing images achieved through physicality. Its flier asked 'What can they possibly hope to gain from their coming together in a city that endlessly tears itself apart?' *Convictions* perhaps suggests a way forward, both in its theatricality and its ethos – expressed by actor Vincent Higgins,

> The production was of now. It was an Irish, Northern Irish, Ulster, nine and six county piece. It was never claimed by one or the other. It is written that way in the Good Friday Agreement, nearly in stone.
>
> (*Arts*, December 2000 p. 22)

Three events in mid-summer 2001 suggest changes in the political climate. First, the Republic's referendum vetoed further enlargement of the EEC. Second, Ulster Unionist David Trimble resigned as First Minister of the Stormont Assembly in protest at slow movement towards arms decommissioning (01 July 2001). Third, further support for the Agreement and Peace Process may be eroded by the increased numbers of DUP and Sinn Fein representatives due to the British elections.

Meanwhile, new work as different as McPherson's interwoven monologues in *Port Authority* (Dublin Gate/New Ambassadors, London) or *Rap Eire* (Project Arts, Dublin) about the duping of an American visitor – performed by actor-satirist Arthur Riordan with supporting cast and hip-hop DJ – ushered in 2001. Amongst this year's premieres were to be Carr's *Ariel* (Project Arts), Walsh's *Bedbound* (Corcadora), Roche's *On Such As We* (Tricycle), Geraldine Aron's *My Brilliant Divorce* (Druid) and McDonagh's *The Lieutenant of Inishmore.* The latter, produced by the Royal Shakespeare Company after rejection by the Royal Court and London's National Theatre caused a sensation. A savage, blood-soaked comedy, it 'satirises the private puritanism, sexual vanity and search for historical sanctions of the paramilitaries'(Billington, *The Guardian* 14 June 2001), through INLA leader 'mad Padraic's excessive grief and revenge for his dead cat'. In contrast was Roddy Doyle's *Guess Who'se Coming to Dinner* (St. Andrew's Lane, Dublin), a liberal exploration of domestic racism in an Irish context, inspired by Stanley

Kramer's 1967 film. The Gate produced three short new plays: Friel's, *The Yalta Game*, Neil Jordan's *In White Horses* and McPherson's *Come On Over*. At two successive Edinburgh Festivals, the Corn Exchange triumphed, first with the *Car Show*, four plays in cars with the audience in the back seats, and secondly with Michael West's *Foley*, a monologue spoken by a Protestant living in the Republic. A revival of Friel's *Faith Healer* (Almeida, November, 2001) prefigured new productions in early 2002 of McGuinness' *Factory Girls* and Murphy's *Conversations on a Homecoming* (both Lyric, Belfast), Roche's *The Cavalcaders* (Tricycle) and Burke-Kennedy's *Women in Arms* (Civic, Dublin). BBC 2 showed Mitchell's film version of his stage play *As the Beast Sleeps*. New plays early in 2002 included Mercier's *Diarmuid and Grainne* (Passion Machine at the Olympia, Dublin), Farrell's *Lovers at Versailles* (Abbey), two Chekhovian plays by Friel (Gate), *Scenes from a Water Cooler* (St Andrews Lane) by Paul Meade and David Parnell, Eamonn Sweeney's *Bruen's Twist* (Everyman Palace, Cork) and Max Hafler's *Grand* (Limerick). Barry's *Hinterland* (Abbey/Out of Joint) caused a stir with its satirical exploration of political corruption. Fiachra Gibbons in her review (*The Guardian* 16 Feb. 2002) cites the artistic Director Ben Barnes' anxiety about the freezing of the Abbey grant. It is to be hoped that future funding problems for the arts in general will not restrict the veritable cornucopia of drama and performance selectively cited above.

Although it is impossible to do justice to all the new writers and the wealth of groups and venues within the scope of this book, it is clear that on both sides of the border, key companies now share in visual and physical experimentation, which pushes against the highly verbal and realist traditions of the past. Even where new writing does not demand intense physicality, the issue of cultural identity is either subtextually present or imprinted within the performing body. These two formal and thematic elements, which have been considered to be characteristics of postcolonial discourse, can in the current context be linked to a movement beyond both the past and the shores of Ireland as a whole, through the international relationships formed by creators of contemporary Irish drama. Just as the population has become more culturally and ethnically diverse, economic relationships with the European community, multinational companies and American influences have contributed further to a destabilisation of previous, more simplistic notions of authentic Irish identity. The image of the rural hearthside has been superceded by the heterotopic site within which the endless multiple possibilities for flexible cultural identities can continue to be celebrated.

Bibliography

Primary Sources

i) Anthologies

Bolger, Dermot (ed.) *Greatest Hits*, Ireland, New Island Books, London, Nick Hern Books, 1997.

——O'Flatharta, Antoine, *Blood Guilty.*

——Dowling, Clare, *The Marlborough Man.*

——McLaughlin, Thomas, *Greatest Hits.*

——MacKenna, John, *Faint Voices.*

Burke, Siobhan (ed.), *Rough Magic First Plays*, Dublin, New Island Books, 1999.

Boyd, Pom, *Down Onto Blue.*

——Hughes, Declan, *I Can't Get Started.*

——Meehan, Paula, *Mrs. Sweeney*

——Moxley, Gina, *Danti-Dan.*

——O'Kelly, Donal, *The Dogs.*

——Riordan, Arthur, *Hidden Charges.*

Fairleigh, John (ed.) *Far From the Land*, London, Methuen, 1998.

——Carville, Daragh, *Language Roulette.*

——McCabe, Patrick, *Frank Pig Says Hullo.*

——Morrison, Conall, *Hard to Believe.*

——O'Kelly, Donal, *Bat the Father, Rabbit the Son.*

——Walsh, Enda, *Disco Pigs.*

——Woods, Vincent, *At the Black Pig's Dyke.*

Grant, David (ed.), *The Crack in the Emerald*, 2nd edn, London, Nick Hern Books, 1994.

——Bolger, Dermot, *The Lament for Arthur Cleary.*

——Carr, Marina, *Low in the Dark.*

——Harding, Michael, *The Misogynist.*

——Jones, Marie, The Hamster Wheel.

Fitz-Simon, Christopher, & Sternlicht, Sandford (eds), *New Plays from the Abbey Theatre*, Syracuse, New York, 1996.

——Donnelly, Neil, *The Duty Master.*

——Harding, Michael, *Hubert Murray's Widow.*

——Mac Intyre, Tom, *Sheep's Milk on the Boil.*

——O'Kelly, Donal, *Asylum! Asylum!*

——Williams, Niall, *A Little Like Paradise.*

McGuinness, Frank (ed.) *The Dazzling Dark*, London, Faber & Faber, 1996.

——Carr, Marina, *Portia Coughlan.*

——MacIntyre, Tom, *Good Evening, Mr. Collins.*

——Moxley, Gina, *Danti-Dan.*

——Murphy, Jimmy, *A Picture of Paradise.*

ii) Playtexts

Barry, Sebastian, (edn. Introduced by Barry), *Prayers of Sherkin, Boss Grady's Boys*, London, Methuen Drama, 1991.

——*The Steward of Christendom*, London, Methuen/Royal Court, 1995a).

——(edn Introduced by F. O'Toole),*The Only True History of Lizzie Finn, The Steward of Christendom, White Woman Street*, London, Methuen, 1995b).

——*Our Lady of Sligo*, London, National Theatre/Methuen, 1998.

——*Plays 1*, Methuen, 1998.

Bolger, Dermot, *Dublin Quartet: The Lament for Arthur Cleary, The Holy Ground, Blinded by the Light, One Last*

——*White Horse*, London, Penguin, 1992

——*A Dublin Bloom*, Dublin/London, New Island/Nick Hern, 1995.

——*April Bright, Blinded by the Light*, Dublin/London, New Island Books/Nick Hern, 1997.

——*Plays 1: The Lament for Arthur Cleary, In High Germany, The Holy Ground, Blinded by the Light*, London, Methuen, 2000.

Burke-Kennedy, Mary Elizabeth, (unpublished script), *Women in Arms*, 1993.

Carr, Marina, *Portia Coughlan*, London, Royal Court/Faber & Faber, 1996.

——(reprint) *The Mai*, Loughcrew, Ireland, Gallery Press, 1997.

——*By the Bog of Cats*, Loughcrew, Ireland, Gallery Press, 1998.

——*On Raftery's Hill*, London, Druid/Royal Court/Faber, 2000.

——*Ariel*, Loughcrew, Ireland, The Gallery Press, forthcoming.

—see Anthology (ed.) Grant, D. for *Low in the Dark*.

Croghan, Declan, *Paddy Irishman, Paddy Englishman, Paddy [...] ?* London, Birmingham Repertory Theatre/Faber StageScripts, 1999.

Devlin, Ann, (2nd edn), *Ourselves Alone*, London, Faber & Faber, 1990. *After Easter*, London, Faber & Faber, 1994.

Farrell, Bernard, *Happy Birthday, Dear Alice, & Stella by Starlight*, Cork, Dublin, 1997.

——*Kevin's Bed*, Dublin, Mercier Press. 1999.

Friel, Brian, *Selected Plays: Philadephia Here I Come!, The Freedom of the City, Living Quarters, Aristocrats, Faith Healer, Translations*, London, Faber & Faber, 1984.

——(2nd edn), *Volunteers*, Loughcrew, Ireland, Gallery Press 1989.

——*Dancing at Lughnasa*, London, Faber & Faber, 1990.

Gregory, Lady Augusta, *Selected Plays Lady Gregory*, Gerrard's cross, UK, Colin Smythe, 1983.

Harding, Michael, see Anthologies (ed.) Grant, D., *The Misogynist*, and (ed.) Fitz-Simon, C., *Hubert Murray's Widow*.

Hughes, Declan, *Plays 1: Digging for Fire, New Morning, Halloween Night, Love & a Bottle*. London, Methuen, 1998.

——See also Anthology (ed.) Burke, S, *I Can't Get Started*.

Hutchinson, Ron, (revised edn.), *Rat in the Skull*, London, Royal Court/Methuen, 1995.

Johnson, Jennifer, *Three Monologues: Twinkletoes, Mustn't Forget High Noon, Christine*, Belfast, Lagan Press, 1992.

Jones, Marie, *A Night in November*, Dublin/London, New Island/Nick Hern, 1995.

——*Women on the Verge of HRT*, London, Samuel French, 1999.

——*Stones in His Pockets, A Night in November*, London, Nick Hern, 2000.

——See Anthology (ed.) Grant, D, *The Hamster Wheel*.

Keane, John. B. (revised texts, Barnes, B. (ed.)), *Three Plays; Sive, The Field, Big Maggie*, Dublin, Mercier Press, 1990.

Kilroy, Thomas, (2nd edn.) *Double Cross,* Loughcrew, Ireland, Gallery Press, 1994. (new edn.).

——*Talbot's Box,* Loughcrew, Ireland, Gallery Press, 1997.

——*The Secret Fall Of Constance Wilde,* Loughcrew, Ireland, Gallery Press, 1997.

Leonard, Hugh, *Selected plays of Hugh Leonard,* Gerrard's Cross, UK, Colin Smythe, 1992.

Lynch, Martin, *Three Plays: Dockers, The Interrogation of Ambrose Fogarty, Pictures of Tomorrow,* Belfast, Lagan Press, 1996.

MacIntyre, Tom, *The Great Hunger, Poem into Play, (with text of Kavanagh's poem)* Gigginstown, Ireland, The Lilliput Press, 1988.

——See Anthologies (ed.) Fitzsimon, C., *Sheep's Milk on the Boil,* and (ed.) McGuinness, F., *Good Evening, Mr.Collins.*

McDonagh, Martin, *The Beauty Queen of Leenane,*London, Druid/Royal Court/Methuen, 1996.

——*The Lonesome West,* London, Druid/Royal Court/Methuen, 1997. *A Skull in Connemara,* London, Druid/Royal Court/Methuen, 1997.

——*The Cripple of Inismaan.* London, Methuen, 1997.

McGuinness, Frank, *Observe the Sons of Ulster Marching Towards the Somme,* London, Faber & Faber, 1986.

——*Mary & Lizzie,* London, Faber & Faber, 1989.

——*Plays 1: The Factory Girls, Observe the Sons of Ulster Marching Towards the Somme, Innocence, Carthaginians, Baglady,* London, Faber & Faber, 1996.

——*Mutabilitie,* London, Faber & Faber, 1997.

——*Dolly West's Kitchen,* London, Faber & Faber, 1999.

McPherson, Conor, *This Lime Tree Bower, Rum & Vodka, The Good Thief,* Dublin, London, New Island/Nick Hern, Bush Theatre, 1996.

——*The Weir,* London, Nick Hern/Royal Court, 1997.

——*St. Nicholas* included in *Four Plays,* London, Nick Hern, 1999.

——*Dublin Carol,* London, Nick Hern/Royal Court, 2000.

Meehan, Paula, see Anthology, (ed.) Burke, S. *Mrs. Sweeney,* 1999.

Mercier, Paul, *Down the Line,* with Bourke, Ken, *The Hunt for Red Willie,* Dublin, Methuen/Abbey Theatre, 2000.

Mitchell, Gary, (unpublished script), *A Little World of Our Own.* Now available, with *Tearing the Loom,* London, Nick Hern, 1998.

——*Trust,* London, Nick Hern/Royal Court, 1999.

——*The Force of Change,* London, Nick Hern/Royal Court, 2000.

Moxley, Gina, *Doghouse* in (ed.) Drake, N. *New Connections,* London, Faber & Faber 1997.

——see Anthologies (ed.), Burke, S., & (ed.) McGuinness, F., *Danti-Dan.*

Murphy, Jimmy, *Brothers of the Brush,* London, Oberon Books, 1995.

——see Anthology (ed.) MacGuinness, F., *A Picture of Paradise.*

Murphy, Tom, *Plays 1: Famine, The Patriot Game, The Blue Macushla,* London, Methuen, 1992.

——*Plays 2: Conversations on a Homecoming, Bailegangaire, A Thief of Christmas,* London, Methuen. 1993.

——*Plays 3: The Morning After Optimism, The Sanctuary Lamp, The Gigli Concert,* London, Methuen 1994.

——*The Wake,* London, Methuen 1999.

O'Casey, Sean, (ed. Ayling, Ronald) *Seven Plays,* MacMillan, 1985.

O'Connor, Joseph, *Red Roses & Petrol,* London, Methuen, 1995.

O'Kelly, Donal, See Anthologies, (ed. Burke, S.) *The Dogs,* (ed. Farleigh, John), *Bat the Father, Rabbit the Son.*

Parker, Stewart, (reprint) *Three Plays for Ireland: Northern Star, Heavenly Bodies, Pentecost,* London, Oberon Books, 1995.

——*Catchpenny Twist*, Dublin, Gallery Press, 1980.

Reid, Christina, *Joyriders & Did You Hear the One About the Irishman?* Oxford, Heinemann Educational. 1993.

——*Plays 1: Tea in a China Cup, Did You Hear the One About the Irishman? Joyriders,The Belle of Belfast City, My Name Shall I Tell You My Name? Clowns*, London, Methuen, 1997.

Reid, Graham, *The Billy Plays*, London, Faber & Faber, 1984.

Roche, Billy, (2nd edn.) *The Wexford Trilogy: A Handful of Stars, Poor Beast in the Rain, Belfry*, London, Nick Hern, 1993.

Stembridge, Gerard, *The Gay Detective*, Dublin/London, New Island, Nick Hern, 1996.

Synge, John, M., ((ed.) Saddlemyer, A.) *Complete Plays*, Oxford, Oxford University Press, 1995.

Walsh, Enda, *Disco Pigs & Sucking Dublin*, London, 1997.

Woods, Vincent, See Anthology (ed.) Fairleigh, J., *At the Black Pig's Dyke*.

Yeats, William B. (reprint) *Selected Plays*, London, MacMillan, 1992.

iii) Other

Constitution of the Irish State, Dublin, 1937.

Friel, Brian, (Christopher Murray, ed), *Brian Friel, Essays, Diaries, Interviews*, London, New York, Faber & Faber, 1999.

Gregory, Lady, (Coole edn.) *Our Irish Theatre*, Gerards Cross, Colin Smythe, 1973.

Kavanagh, Patrick, (pbk reprint) *Tarry Flynn*, London, Penguin, 1978.

Kinsella, Thomas, (trans), *The Tain*, (2nd edn.). Oxford, Oxford University Press, 1970.

McCabe, Patrick, *The Dead School*, London, Picador, 1996.

O'Casey, Sean, *The Green Crow*, London, W.H. Allen, 1985.

Spenser, Edmund, (Hadfield, A, Maley, W. eds) *A View of the State of Ireland*, Oxford, USA, Blackwell. 1997.

Synge, J. M., ((ed.) Price, A.) *Collected Works, Vol. 2 Prose*, Oxford, Oxford University Press 1966.

Yeats, W. B. (ed.) *The Poems of Edmund Spenser* including *Mutabilitie*, London, Caxton, old undated volume.

Secondary Sources

Artaud, Antonin, (reprint), *The Theatre & its Double*, London, John Calder, 1981.

Ashcroft, Bill, Griffith, Gareth, & Tiffin, Helen, *The Post-colonial Studies Reader*, (3rd reprint), London, USA, 1997.

Aston, Elaine, & Savona, George, *Theatre as Sign System*, London, USA, Routledge,
1991.

Baudrillard, Jean, *Symbolic Exchange & Death*, trans. I. H. Grant, (reprint), London, California, Sage, 1995

Bakhtin (ed Morris, Pam) *The Bakhtin Reader*, London, Edward Arnold, 1994.

Barton, Ruth, 'The Ballykissangelization of Ireland', pp. 413–26 in *Historical Journal of Film, Radio & Television'*, London, 2000.

Beale, Jenny, *Women in Ireland: Voices of Change*, Dublin, Gill & Macmillan, 1986.

Belsey, Catherine, *Critical Practice*, London, Methuen, New Accents, 1980.

Benjamin, Walter, 'The Author as Producer' in *Understanding Brecht*, London, New Left Books, 1973.

Bennett, Susan, *Theatre Audiences; a Theory of Reception*, London, USA, Routledge, 1990.

Bhabha, Homi K., *The Location of Culture* (3rd reprint), London, USA, Routledge, 1997.

Bharuca, Rustom, (2nd edn.)*Theatre & the World: Performance & the Politics of Culture*, London, New York, Routledge, 1993.

Birringer, Johannes, *Theatre, Theory & Postmodernism*, USA, Indiana University Press, 1991.

Bibliography

Bradley, Anthony, & Valuilis, Maryann G., *Gender & Sexuality in Modern Ireland*, Amherst, USA, University of Massachusetts Press, 1997.

Brecht, Bertholt, (trans. John Willett), (reprint) *Brecht on Theatre*, London, Methuen, 1982.

Brett, David, *The Construction of Heritage*, Cork, Ireland, Cork University Press, 1996.

Brewster, Scott, Crossman, Virginia, et al. (eds) *Ireland in Proximity, History, Gender, Space*, London, New York, Routledge, 1999.

Brook, Peter, *The Empty Space*, London, USA, Penguin, 1968.

Cairns, David & Richards, Shaun, (reprint) *Writing Ireland*, Manchester, New York, Manchester University Press, 1996)

Chambers I. & Curtin, L., *The Post-Colonial Questions*, London, Routledge, 1996.

Coulter, Colin, *Contemporary Northern Irish Society*, London, USA, Pluto Press, 1999.

Counsell, Colin, *Signs of Performance*, London, USA, Routledge, 1996.

Crow, Brian, & Banfield, Chris, *An Introduction to Post-colonial Theatre*, Cambridge, New York, Cambridge University Press, 1996.

Curtin, Chris, Jackson, Pauline, O'Connor, Barbara, (eds), *Gender in Irish Society*, Galway, University of Galway, 1987.

Deane, Seamus, *Celtic Revivals*, (pprbk edn.), Winston Salem, USA, Wake Forest University Press/Faber, 1987.

Douglas, Mary, *Purity and Danger*, London, USA, RKP, 1966.

Eagleton, Terry, *Heathcliff and the Great Hunger*, London, New York, Verso, 1995.

Elam, Keir, *The Semiotics of Theatre & Drama*, London, New York, Methuen New Accents, 1980.

Etherton, Michael, *Contemporary Irish Dramatists*, London, Macmillan, 1989.

Flannery, James W. *W.B. Yeats & the Idea of a Theatre*, (pbk, later edn), Newhaven, London, Yale University Press, 1989.

Foster, Roy, *Modern Ireland, 1600–1972*, (1st Penguin edn), London, USA, Penguin, 1989.

——*Paddy & Mr. Punch*, (1st Penguin edn.) London, USA, Penguin, 1995.

Foucault, Michel, (trans.A. Sheridan),*Discipline and Punish*, London, New York, Penguin, 1997.

——(trans. Martin et al.) *The Technologies of the Self*, London, MacMillan, 1988.

——'Of Other Spaces', pp. 22–7, in *Diacritics, Volume 16, Part 1*, Spring, 1986.

Freud, Sigmund, ((ed.) Albert Dickson), *Volume 14: Art & Literature*, London, New York, Penguin, 1985.

Gibbons, Luke, (ed.) *Transformations in Irish Culture*, Cork, Cork University Press/Field Day, Cork, Ireland, 1996.

Gilbert, Helen, & Tompkins, Joanne, *Post-colonial Drama: Theory, Practice, Politics*, London, New York, Routledge, 1996.

Graham, Colin, & Kirkland, Richard, *Ireland & Cultural Theory: the Mechanics of Authenticity*,
Britain, MacMillan; New York, St. Martin's Press, 1999.

Grene, Nicholas, *The Politics of Irish Drama: Plays in Context from Boucicault to Friel*, Cambridge, New York, Cambridge University Press, 1999.

Griffiths, Trevor R., & Llewellyn-Jones, Margaret, (eds), *British & Irish Woman Dramatists Since 1958*, Buckingham, UK, Bristol USA, Open University Press, 1993.

Grosz, Elizabeth, *Space, Time & Perversion: Essays on the Politics of Bodies*, New York, London, Routledge, 1995.

Grotowski, Jerzy, (reprint), *Towards a Poor Theatre*, London, Methuen, 1986.

Hickman, Mary. J. *Religion, Class & Identity: the State, the Catholic Church and the Education of the Irish in Britain*, London, Ashgate, 1995.

Kearney, Richard, *Transitions: Narratives in Modern Irish Culture*, Dublin, Wolfhound, 1988.

PostNationalist Ireland: Politics, Culture, Philosophy, London & New York, Routledge, 1997.
Kenneally, Michael, (ed.), *Irish Literature & Culture*, England, Colin Smythe, 1992.
Kershaw, Baz, *The Politics of Performance*, London, USA, Routledge, 1992, 1999.
Kiberd, Declan, *Field Day Pamphlet Series 2, No 6*, 1984: London, Hutchinson, 1985.
Inventing Ireland, London, Australia, New Zealand, South Africa, Jonathan Cape, 1995.
'Romantic Ireland's Dead & Gone', pp. 12–14, *Times Literary Supplement*, 12th June, 1988.
Kirkland, Richard, *Literature & Culture in Northern Ireland Since 1965*, London New York, Longmans, 1996.
Kristeva, Julia, (trans. Leon Roudiez), *Powers of Horror, An Essay on Abjection*, USA, Columbia, 1992.
(ed.) Toril Moi *The Kristeva Reader*, Oxford, Blackwells, 1986.
Lacan, Jacques, *The Four Fundamental Concepts of Psychoanalysis*, London, Penguin, 1979. *Ecrits* (trans. A. Sheridan), London, Tavistock, 1977.
Levi-Strauss, Claude, (trans. C. Jacobson & B.G. Schoepf), *Structural Anthropology*, England, Penguin, 1972.
Llewellyn-Jones, Margaret, (ed.) *Spectacle, Silence & Subversion: Women's Performance Languages & Strategies; Contemporary Theatre Review Vol. 2 Part 1*, Harwood Academic, Gordon & Breach, Switzerland, 1994.
Lysaght, Patricia, *The Banshee: a Supernatural Irish Death Messenger*, Dublin, Glendale Press, 1986.
Mahoney, Christina H., *Contemporary Irish Literature, Transforming Tradition*,London, MacMillan, 1998.
Maxwell, D.E.S. *Modern Irish Drama 1891–1980*, (reprint), Cambridge University Press, 1988.
McCann, Sean *The Story of the Abbey Theatre*, London, Four Square Books, 1967.
McMullan, Anna, 'Irish Women Playwrights Since 1958,' in Griffiths, T.R. & Llewellyn-Jones, M. *British & Irish Women Playwrights Since 1958*, Buckingham, UK, Open University, 1993.
Mercier, Vivian, *The Irish Comic Tradition*, (British edn, 3rd) London, Souvenir Press, 1991.
Muinzer, Philomena, 'Evacuating the Museum: the Crisis of Playwriting in Ulster', *New Theatre Quarterly, Volume 3, No. 9*, February 1987.
Murray, Christopher, *Twentieth Century Irish Drama, Mirror Up to Nation*, Manchester, New York, 1997.
Nelson, Robin, *TV Drama in Transition*, London, New York, 1997.
Nietzsche, Friedrich, *The Birth of Tragedy*, (trans. F. Golfing), New York, London, Doubleday, 1956.
Nikolakis, Michael, *Staging Identity* unpublished PhD thesis, University of North London, 2001.
O'Brien Johnson, Toni, *Synge, The Mediaeval & the Grotesque*, Gerard's Cross, Smythe, 1982.
O'Brien Johnson, Toni, & Cairns, David, (eds), *Gender in Irish Writing*, Buckingham, UK, Open University, 1991.
O'Grada, Cormac, *The Great Irish Famine*, (1st Cambridge edn), Cambridge, 1995.
O'Toole, Fintan, *Black Hole, Green Card*, Dublin, New Island Books, 1994
——*Tom Murphy: The Politics of Magic*, Dublin, London, New Island/Nick Hern, 1994.
——*The Ex-Isle of Erin*, (reprint) Dublin, New Island Books, 1998.
Paulin, Tom, Field Day Pamphlet Series 1, No 2. 1983: London, Hutchinson, 1985.
Pavis, Patrice, (trans. Loren Kruger) *Theatre at the Crossroads of Culture*, London, New York, 1992.
Pettit, Lance, *Screening Ireland, Film & Television Representation*, Manchester, New York,
Manchester University Press, 2000.
Pine, Richard, *Brian Friel & Ireland's Drama*, London, USA, Canada, 1990.
Richtarik, Marilyn, *Acting Between the Lines, The Field Day Company 1980–1984*, Oxford, Clarendon, 1994.
Roche, Anthony, *Contemporary Irish Drama from Beckett to MacGuinness*, Dublin, Gill & MacMillan, 1994.
Rockett, Kevin, Gibbons, Luke, & Hill, John, *Cinema & Ireland*, London, Routledge, 1987.
Ryan, Louise. & Gray, Breda 'Gendered Constructions of Irishness: Stagnation or Change in Irish Society since Independence', in Stern-Gillet, Slawek, Tadeusz et al., *Culture & Identity: Selected Aspects and Approaches*, Katowice, Poland, Wydawnicto Uniwersytetu Slaskiego, 1996.

Said, Edward, *Field Day Pamplet Series 5, No. 15,* 1988: in (ed.) Deane, S., *Nationalism, Colonialism & Literature,* USA, University of Minnesota, 1989.

Sales, Rosemary, *Women Divided, Gender, Religion & Politics in Northern Ireland,* London, USA, Routledge, 1997.

Smyth, Aibhe, *Irish Women's Studies Reader,* Dublin, Attic Press, 1993.

——'Feminism: Personal, Political, Unqualified (or Ex-Colonised Girls Know More', pp. 37–54, in *Irish Journal of Feminist Studies, Volume 2 issue 1,* Cork University Press, 1997.

Smyth, Daragh, *A Guide to Irish Mythology,* (2nd edn.), Dublin, USA, Irish Academic Press, 1996.

Smyth, Gerry, *Decolonisation & Criticism: the Construction of Irish Literature,* London, Pluto Press, 1998.

Stallybrass, Peter & White, Allon, *The Politics and Poetics of Transgression,* London, Methuen, 1986.

Sternlicht, Sanford, *A Reader's Guide to Modern Irish Drama,* New York, Syracuse University Press, 1998.

Turner, Victor, *From Ritual to Theatre,* New York, PAJ Publications, 1982.

Ward, Margaret, *The Missing Sex, Putting Women into Irish History,* Dublin, Attic Press, 1991.

Woodham-Smith, C. *The Great Hunger 1845–1859,* London, Penguin, 1962.

Worth, Katherine, *The Irish Drama of Europe from Yeats to Beckett,* (pbk), London, The Athlone Press, 1986.

ii) Journals

Irish Journal of Feminist Studies, Volume 2, Issue 1, Aspects of Irish Feminism, Cork University Press, July 1997.

Irish Review, No. 22 Theatre Business, Cork University Press, Summer 1998.

Time Line

NB. This selective list of key events does not include all plays discussed in the book.

DATE	Political events	Theatrical events
1171	Ireland becomes a 'lordship' of English crown after invasion by Henry II of England, sanctioned by the Pope. English law applied to 'the Pale' around Dublin	
1541	Henry VIII claims title King of Ireland Tudor 'plantation' of English gentry as landlords in Ireland	(See McGuinness' *Mutabilitie* 1998 about Elizabethan Ireland)
1601	Hugh O'Neill's rebellion falters after defeat at battle of Kinsale	
1649	As Governor of Ireland, Oliver Cromwell causes massacre at Drogheda	
1660	Restoration of Charles II in England	
1688	Protestant William of Orange is invited to be King of England	
1690	Battle of the Boyne – William's Protestant army defeats Irish rebels supporting Catholic James II, deposed from English throne	
1791	United Irishmen formed in Belfast & Dublin	
1795	Founding of the (Protestant) Orange Order	
1800	Act of Union passed by (English) Parliament	
1833	Ordnance Survey of Ireland by British Army Engineering Corps	(See Friel *Translations* 1980)
1834	Daniel O'Connell introduces debate on repeal of Union	
1838	English Poor Law extended to Ireland	
1845–1859	The Great Famine – emigration increase	(See Murphy *Famine* 1968)
1852	First Saint Patrick's Day Parade in New York	
1860		Boucicault *The Colleen Bawn*
1862	Harland & Wolff shipyard in Belfast	
1864		Boucicault *The Shaughraun*
1867	Fenian disturbances in England & Ireland.	
1869	Disestablishment of Church of Ireland	
1870	Home Rule movement founded	
1880	Parnell leader of Irish Parliamentary Party	
1886	Gladstone's Home Rule Bill defeated	

1887		Antoine's Theatre Libre, Paris
1890		Ibsen's *Ghosts* in Paris
1892		Shaw writes *Widower's Houses*
1893	Second Home Rule Bill, passed by Commons rejected by Lords	
1897		Moscow Arts Theatre formed
1899		Establishment of Irish Literary Theatre.
1902		Yeats' *Cathleen ni Houlihan*
1903		Irish National Theatre Society founded
1904		Abbey theatre opened
1905	Sinn Fein established	
1907	Jim Larkin – Union leader, organises dock strike in Belfast	Synge's *Playboy of the Western World*
1911	Irish Women's Suffrage movement	
1912	Third Home Rule Bill, introduced – but undergoes various challenges Ulster covenant against Home Rule	
1913	Irish Transport & General Worker's Dublin 'Lock out', led by Larkin, then James Connolly. Formation of Citizen Army, Foundation of Irish Volunteers.	(See Kilroy *Talbot's Box* 1977)
1914	Home Rule eventually passed but suspended due to 1914–1918 First World War	(See McGuinness *Behold the Sons of Ulster* 1985)
1916	Easter Rising; execution of leaders	
1919	War of Independence begins, Irish groups versus British troops	
1920	Government of Ireland Act gives separate northern and southern Irish parliaments – both within Britain. (British) Black & Tan soldiers perpetrate first Bloody Sunday	
1921	Anglo Irish Treaty signed	
1922	Michael Collins involved in negotiations re creation of Irish Free State, leaving (6 counties) Northern Ireland as part of Britain. Civil War	(See Barry *Steward of Christendom* 1995)
1923	Civil War ends	O'Casey *Shadow of a Gunman*
1926		O'Casey *Plough & the Stars*
1929		Johnston *The Old Lady Says No!*
1930		Gate opens its own premises
1934		Flaherty's film *Man of Aran*
1937	De Valera's Fianna Fail government creates New Irish Constitution, names country as Eire, asserts right to claim Northern Ireland.	
1938		Artaud writes *Theatre & Its Double*

Time Line

1938–45	Second World War – Eire remains neutral, Northern Ireland involved	(See McGuinness *Dolly West's Kitchen* 1999)
1949	Ireland leaves British Commonwealth, becomes a Republic	Brecht forms Berliner Ensemble.
1951–1961	The highest emigration rates yet from Ireland	
1952		Ford's film *The Quiet Man*
1953		Pike theatre established
1954		Behan's *The Quare Fellow*
1955		Beckett's *Waiting for Godot*
1956–1962	IRA Border campaign	
1958		Behan *The Hostage*
1962		(See Lynch *Dockers* 1981 set in Belfast)
1963		Brook's Theatre of Cruelty season in London.
1964		Friel *Philadelphia Here I Come*
1965		Keane *The Field* (stage)
1966		New Abbey Theatre opens.
1968		Murphy's *Famine*. Grotowski writes *Towards a Poor Theatre* Lyric Theatre Belfast gains building
1969	Civil Rights campaign in Northern Ireland marks start of violence & 'the Troubles'	
1972	Bloody Sunday, Derry – British troops shoot civilians. Period of Direct Rule of Northern Ireland from Westminster	
1973	Ireland joins the EEC – which will influence its move towards a 'Celtic Tiger' Economy Sunningdale agreement sets up Power-sharing Executive on Northern Ireland. From early 1970's decline of the North's manufacturing base leads to huge unemployment during 1980's & early 1990's	
1974	Power-sharing Executive brought down through Ulster Worker's Council action	(see Parker's *Pentecost* 1988)
1975		Druid Theatre, Galway formed Friel's *Volunteers*
1977		Kilroy *Talbot's Box*
1980		Field Day Theatre Company – Friel's *Translations* is first play
1981	Hunger Strikes in Northern Ireland	
1982		Lynch *The Interrogation of Ambrose Fogarty*
1982–4		Reid, G. *The Billy Plays*
1983	Republic has referendum on divorce	Reid, C. *Tea in a China Cup* Charabanc formed
1984		Hutchinson *Rat in the Skull*

		Rough Magic receives funding Passion Machine's first production
1985	Hillsborough Anglo-Irish Agreement attempts to pave way for devolved government More liberal Amendment to 1979 Health (Family Planning) Act in Republic	Murphy *Bailegangaire* McGuinness *Observe the Sons of Ulster Marching [...]* Reid, C. *Did You Hear the One [...]* Devlin *Ourselves Alone*
1986		Barry *Boss Grady's Boys* MacIntyre *The Great Hunger*
1987	Tax incentives for creative writers, filmmakers etc. introduced in Ireland.	Parker *Pentecost*
1988	60,000 migrants leave Republic	Burke-Kennedy *Women at Arms* Tinderbox formed.
1988–1991		Roche *Wexford Trilogy* Macnas founded
1989		Bolger *A Lament for Arthur Cleary*
1990	Mary Robinson President of Republic	Barry *Prayers of Sherkin* Bolger *In High Germany* Friel *Dancing at Lughnasa*
1991		Dubbel Joint formed Dock Ward Story Murphy's *The Patriot Game*
1992	Republic referenda on divorce & abortion	Barry *White Woman Street*
1993	Secret talks between British & Irish Prime Ministers & IRA sign Downing Street Declaration – Britain concedes has no strategic interest in Northern Ireland Legalisation of homosexuality in Republic	Murphy, J. *Brothers of the Brush* Pan Pan, Barabbas founded
1994	Ceasefire by IRA and Loyalists Economic improvements begin in North.	(See Reid, C. *Clowns* 1996 & Carville *Language Roulette* 1996) Carr *The Mai* Kabosh formed Lynch *Pictures of Tomorrow* O'Kelly *Asylum, Asylum*
1995	Republic Referendum removes ban on Divorce	Barry *The Steward of Christendom* Devlin *After Easter*
1996	Underbelly of Celtic Tiger revealed in urban plays:	Murphy, J. *A Picture of Paradise* Mercier *Buddleia,*
	rural plays:	Carr *Portia Coughlan* McDonagh *The Beauty Queen of Leenane* McDonagh *The Cripple of Inismaan*
	other:	Jones *Women on the Verge of HRT* Stembridge *The Gay Detective*
1997		Barabbas *White Headed Boy*

		Hughes *Halloween Night* Jones *One Night in November* Kilroy *The Fall of Constance Wilde* Macnas *Diamonds in the Soil* McDonagh *The Lonesome West* Meehan *Mrs Sweeney* Mitchell *In a Little World of Our Own* Walsh *Disco Pigs*
1998	Good Friday Agreement signed in Belfast – Peace Process involving London, Dublin & Belfast	Barry *Our Lady of Sligo* Carr *By the Bog of Cats* McGuinness *Mutabilitie* McPherson *The Weir* Mercier *Dublin Trilogy* Murphy *The Wake* Woods *At the Black Pig's Dyke*
1999	January: Ireland joins EMU December: Irish Government relinquishes Territorial claim to Northern Ireland Devolved Northern Ireland Cabinet meets for 1st time at Stormont 42,000 migrants enter Republic	Croghan *Paddy Irishman, Paddy Englishman, Paddy ...?* Jones *The Stones in His Pockets* Mitchell *Trust*
2000–1	Ulster: Peace Process continues, some disruption from both sides. British elections increase DUP & Sinn Fein M.P's Decommissioning of weapons still problematic, prompting resignation of Assembly First Minister David Trimble (UU) 1/07/01, further unsettling Peace Process. Republic votes against EEC Enlargement. Economic growth continues both sides of border.	Carr *On Raftery's Hill, Ariel* Kabosh *Kicking Space* Workshop Macnas *Ollie Deasy* McDonagh *Lieutenant of Inishmore* McPherson *Dublin Carol, Port Authority* Mercier *Down the Line, We Ourselves* Mitchell *The Force of Change* Pan Pan International symposium Tinderbox *Convictions*

Index